Electrical and Electronic Principles 3

Electrical and Electronic Principles 3

Rhys Lewis B.Sc Tech, C Eng, MIEE, AMCST

Head of Department of Electronic and Radio Engineering,
Riversdale College of Technology, Liverpool

GRANADA
London Toronto Sydney New York

Granada Publishing Limited – Technical Books Division
Frogmore, St Albans, Herts AL2 2NF
and
36 Golden Square, London W1R 4AH
515 Madison Avenue, New York, NY 10022, USA
117 York Street, Sydney, NSW 2000, Australia
60 International Boulevard, Rexdale,Ontario R9W 6J2, Canada
61 Beach Road, Auckland, New Zealand

British Library Cataloguing in Publication Data
Lewis, Rhys
   Electrical and electronic principles 3.
   1. Electric engineering
   I. Title
   621.3   TK145

ISBN 0 246 11814 8

First published in Great Britain 1983 by Granada Publishing

Printed in Great Britain by William Clowes (Beccles) Limited,
Beccles and London

Granada ®
Granada Publishing ®

Contents

Preface

This is one of a new series of texts for telecommunications, electronics and electrical engineering students studying for the Technician Education Council Courses.

The text covers the unit Electrical and Electronic Principles 3 (U81/742), which was revised in 1981 and increased in length to represent 90 hours teaching. Each chapter represents a module from the unit, as indicated by the general and specific objectives preceding the chapter and subsequent sections. Self-assessment exercises are included complete with marking scheme and there are numerous worked examples including those of a multiple-choice nature.

Author's acknowledgements

I would particularly like to express my gratitude to my wife Dawn, without whose patience, encouragement and practical help, particularly in the typing and preparation of the manuscript, the book would not have reached publication.

Rhys Lewis

Self-assessment exercises in this book

Following each chapter there is a self-assessment exercise which will enable you to test how well you have understood and assimilated the material in the chapter. The layout and marking scheme is similar for each exercise, each one consisting of a number of short-answer or multiple-choice questions followed by long-answer questions. The maximum marks awarded for the short questions are 3 to 5 and for the long ones are 14. Marks are indicated at the side of each question. The maximum for each complete exercise is 100. All questions in each exercise should be attempted.

The time taken for each question and exercise will vary from person to person, of course, but the grade of difficulty of the questions has been carefully chosen so that the maximum time that should be taken is as follows:

Short questions (3 marks) : 3 minutes
Short questions (5 marks) : 5 minutes
Long questions (14 marks) : 30 minutes
The whole exercise : 3 hours

When the whole exercise has been completed and marked (using the model solutions given), a guide to the level of pass is as follows:

40–59 : Pass
60–84 : Pass with merit
85 and above : Pass with distinction

If the total obtained is below 40 it indicates that an insufficient understanding has been obtained and the chapter should be carefully reworked especially in those areas of difficulty indicated by the marks gained per question.

When marking, a high standard should be adopted in those parts of questions which are subjective (although with the degree of detail given in the marking scheme these have been reduced to a minimum). It is better to apply a high standard to your own marking than to deceive yourself that you have a better understanding than you actually have!

1 Circuit Theorems

Topic Area: A

General objective

The expected learning outcome is that the student understands circuit theorems.

Introduction

Electrical and electronic systems are often a mixture of complex circuits and at first sight may appear totally bewildering. Fortunately we can reduce almost all circuits to a much simplified *equivalent circuit* and use well-known circuit theorems to tell us how the circuits behave. The circuit theorems we shall describe and use are Thevenin's theorem, Norton's theorem and the maximum power transfer theorem.

Specific objectives

The expected learning outcome is that the student:
1.1   States Thevenin's theorem.
1.2   Solves simple problems using Thevenin's theorem.

Thevenin's theorem

Thevenin's theorem states that any circuit however complex may be theoretically replaced by a voltage generator in series with an impedance. The generator e.m.f. is the voltage which appears across the output terminals of the circuit which is to be replaced, when any load is removed. The impedance in series with the generator has a value equal to the impedance looking back into the output terminals of the circuit which is to be replaced, when all voltage sources in the circuit are replaced by their internal impedances.

The theorem may be applied to all circuits, a.c. or d.c. At frequencies where reactive components of impedances are negligible we may use the word 'resistance' in place of 'impedance' in the theorem.

Let us examine more closely what the theorem says. Suppose we have a complex circuit in which all the voltage sources are replaced by their internal impedances as shown in fig. 1.1a.

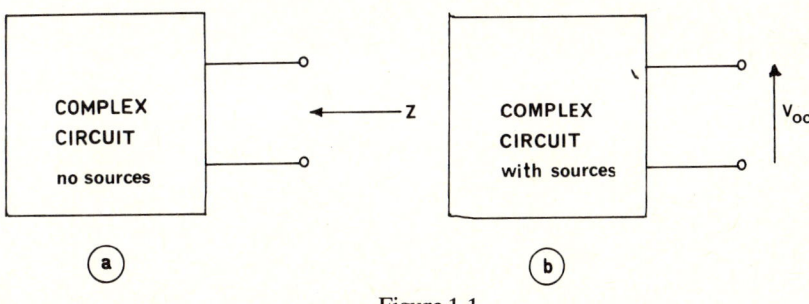

Figure 1.1

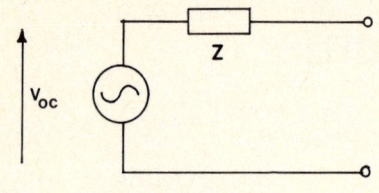

Figure 1.2

The impedance looking back into the output terminals is shown as Z. Suppose when the sources are replaced the voltage appearing across the output terminals is V_{oc} as shown in fig. 1.1b. Thevenin's theorem states that this circuit may be replaced by the simple circuit shown in fig. 1.2.

We see that the simple circuit consists of a voltage generator of e.m.f. V_{oc} in series with an impedance Z.

Now let us apply Thevenin's theorem to a practical circuit, that shown in fig. 1.3.

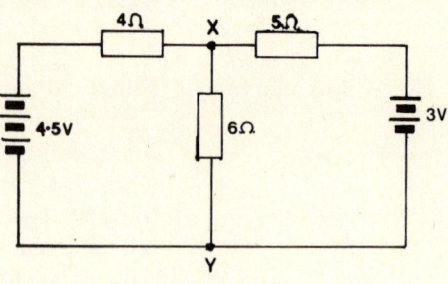

Figure 1.3

Example 1.1   Suppose we wish to find the current flowing in the 6 Ω resistor. This current is flowing due to the combined action of the 4.5 V source in series with a 4 Ω resistor and the 3 V source in series with a 5 Ω resistor. We would like to replace these two sources by a single source using Thevenin's theorem. First, remove the 6 Ω resistor as shown in fig. 1.4a.

The Thevenin equivalent circuit of fig. 1.4a is shown in fig. 1.4b. It consists of a voltage source of e.m.f. E volts, in series with a resistor of resistance R ohms. To find the value of E return to the circuit shown in fig. 1.4a. It is the voltage appearing across XY when the 6 Ω resistor is removed.

When this is done there will be a circulating current through the 4 Ω and 5 Ω resistors in the direction shown. The voltage producing the current is the *difference* between the 4.5 V source and the 3 V source, since one acts so as to cause current flow in a clockwise direction and the other acts so as to cause current flow in an anticlockwise direction. The voltage producing the current flow is thus $(4.5 - 3)$ V, i.e. 1.5 V, and the current, shown as I in the figure, flows clockwise. The total circuit resistance is $(4 + 5)$ ohms, i.e. 9 Ω, and so

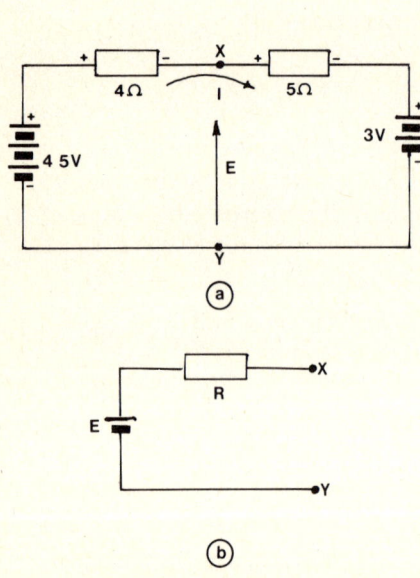

(a)

(b)

Figure 1.4

$$I = \frac{1.5}{9}$$
$$= 0.167 \text{ A}$$

(¹) find the current,

(¹¹)   The voltage across XY is thus
   4.5 V *less* the p.d. across the 4 Ω resistor $4 \cdot 5 - (0 \cdot 167 . 4)$
or
   3 V *plus* the p.d. across the 5 Ω resistor $3 + (0 \cdot 167 . 5)$
(this may be understood more easily by showing the polarity of the voltages using + and − signs as in the figure) i.e.

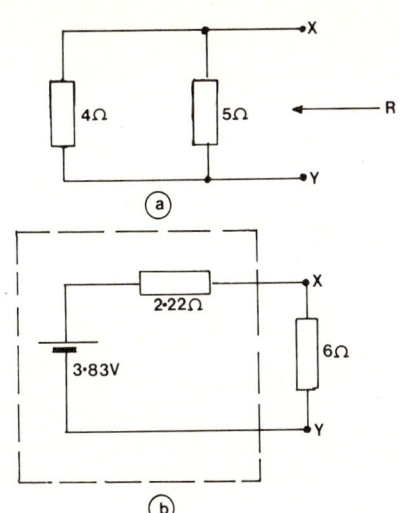

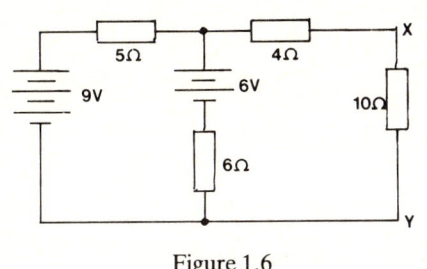

Figure 1.5

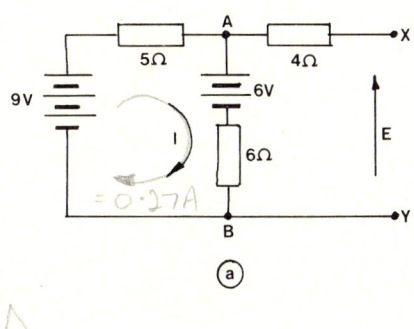

Figure 1.6

$$4.5 - (0.167 \times 4)$$

or

$$3 + (0.167 \times 5)$$

either of which gives the answer 3.83 V. This is the value of E in the Thevenin equivalent circuit shown in fig. 1.4b.

The value of R is obtained by replacing the 4.5 V and 1.5 V sources by their internal resistances. These are not given so that we assume they are zero (or contained in the 4 Ω and 5 Ω resistances, respectively).

Removing the sources then we have a circuit as shown in fig. 1.5a.

The two resistances appear to be in parallel when looking into terminals XY. The equivalent resistance is thus

$$\frac{4 \times 5}{9} \text{ i.e. } 2.22 \ \Omega$$

and the Thevenin equivalent circuit is as shown in fig. 1.5b. Note that the 6 Ω load has now been reconnected.

The calculation to find the current in the 6 Ω resistor is now much simpler. The voltage source is 3.83 V, the total circuit resistance is $(6 + 2.22)\ \Omega$, i.e. 8.22 Ω and the required current equals

$$\frac{3.83}{8.22}\text{A}$$

which gives us 0.466 A.

This example has been explained in some detail. The next two examples are explained in progressively less detail.

Example 1.2 Calculate the p.d. across XY in the circuit shown in fig. 1.6.

First remove the 10 Ω resistor and then replace all the remaining circuit by a Thevenin equivalent circuit as shown in fig. 1.7e.

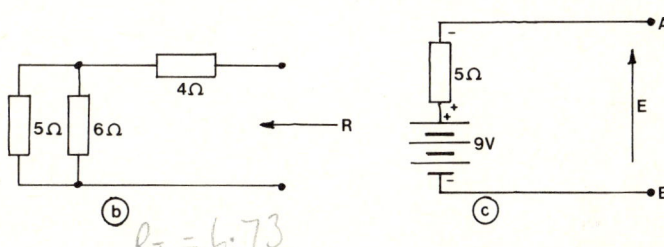

Figure 1.7

When XY is open circuited no current flows in the 4 Ω resistor and E is the voltage appearing across AB. A circulating current, shown as I in fig. 1.7a, will flow due to the combined effect of the 9 V source and the 6 V source through a total resistance of 5 Ω and 6 Ω. The resultant voltage source is $(9 - 6)$ V since the 9 V source acts in opposition to the 6 V source.

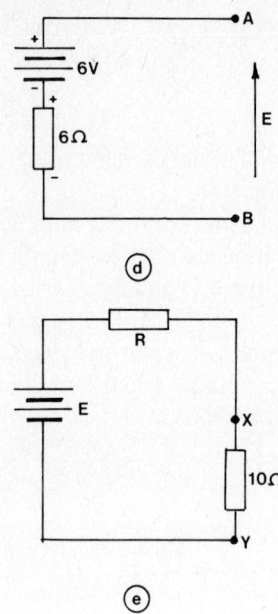

(d)

(e)

Hence,

$$I = \frac{9-6}{5+6}$$

$$= 0.27 \text{ A}$$

and $E = 9 - (0.27 \times 5)$ from fig. 1.7c (or $E = 6 + (0.27 \times 6)$ from fig. 1.7d)

$$E = 7.64 \text{ V}$$

From fig. 1.7b
$$R = 4 + \frac{5 \times 6}{5+6}$$

$$= 6.72 \ \Omega$$

and in the circuit of fig. 1.7e, the current in the 10 Ω resistor is

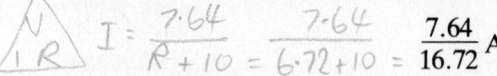

(since the voltage source is 7.64 V and the total resistance is (6.72 + 10) Ω) which equals 0.457 A.

The voltage across the 10 Ω resistor is

$$10 \times 0.457 \text{ i.e. } 4.57 \text{ V}$$

Alternatively, using potential division, the voltage across 10 Ω resistor is

$$\frac{10}{16.72} \times 7.64$$

which again gives us 4.57 V.

Example 1.3   Find the current in the 20 Ω resistor in the circuit shown in fig. 1.8.

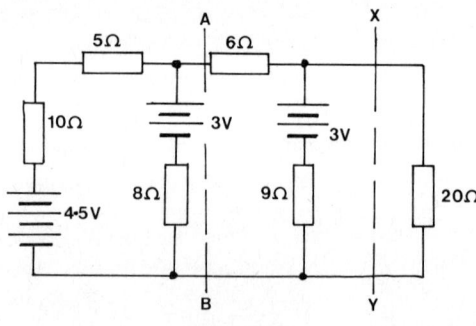

Figure 1.8

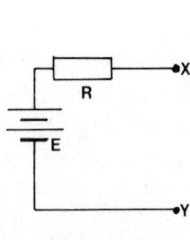

Figure 1.9

The circuit to the left of XY is to be replaced by a Thevenin equivalent circuit as shown in fig. 1.9.

The circuit to the left of XY is quite complex consisting as it does of two meshes. The mesh to the left of AB however may be replaced using Thevenin's theorem as follows:

Effective e.m.f. in the mesh is $4.5 - 3$ i.e. 1.5 V

mesh resistance is $10 + 5 + 8$ i.e. 23 Ω

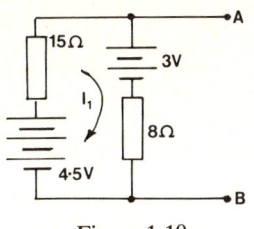

Figure 1.10

and the circulating current

$$I_1 = \frac{1.5}{23}$$
$$= 0.065 \text{ A}$$

The voltage across AB when all the circuit to the right is removed, as shown in fig. 1.10, is then

$$4.5 - 15 \times 0.065 \text{ V}$$
$$(\text{or } 3 + 8 \times 0.065)$$

which equals 3.52 V, and the Thevenin circuit resistance is the parallel combination of 15 Ω and 8 Ω,

i.e. $$\frac{15 \times 8}{23}$$

which is 5.22 Ω.

We may now redraw the circuit of the problem (fig. 1.8) replacing the circuit to the left of XY by a simplified circuit as in fig. 1.11. We will then repeat the process.

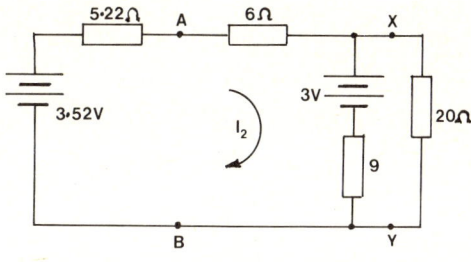

Figure 1.11

Removing the 20 Ω resistor across XY, the circuit current

$$I_2 = \frac{3.52 - 3}{5.22 + 6 + 9}$$
$$= 0.026$$

and the open circuit voltage across XY shown as E in the overall Thevenin equivalent circuit of fig. 1.9 is

$$3.52 - 0.026\,(5.22 + 6) \text{ V}$$
$$(\text{or } 3 + 9 \times 0.026 \text{ V})$$

which equals 3.23 V.

The resistance R in the overall equivalent circuit consists of (5.22 + 6) Ω in parallel with 9 Ω

i.e. $$\frac{11.22 \times 9}{11.22 + 9}$$

which equals 4.99 Ω, and the initial circuit may be redrawn as in fig. 1.12.

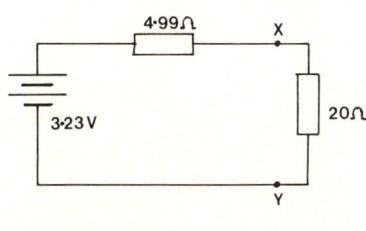

Figure 1.12

The current in the 20 Ω resistor is

$$\frac{3.23}{4.99 + 20} \text{ i.e. } 0.129 \text{ A}$$

EXERCISE 1.1

1. Calculate the current in the 5 Ω resistor in the circuit of fig. 1.13 using Thevenin's theorem.

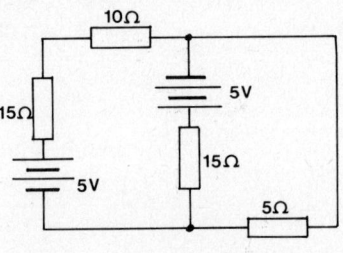

Figure 1.13

2. Calculate the voltage across XY in the circuit shown in fig. 1.14 using Thevenin's theorem.

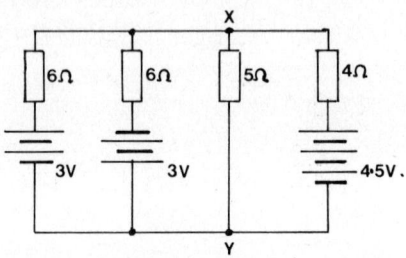

Figure 1.14

3. Determine the Thevenin equivalent circuit of the circuit shown in fig. 1.15.

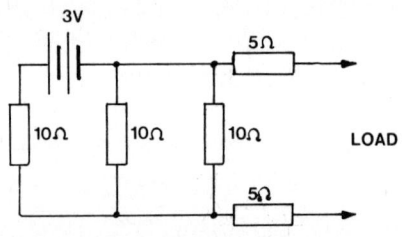

Figure 1.15

4. The Thevenin equivalent circuit of the circuit shown in fig. 1.16 has a resistance of value 10 Ω. What is the value of *R*?

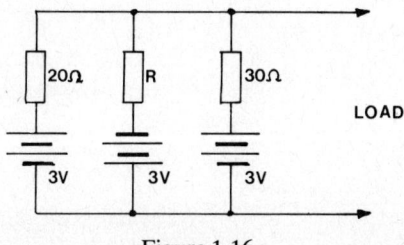

Figure 1.16

5. The circuit shown in fig. 1.17a is the Thevenin equivalent of the circuit to the left of AB in fig. 1.17b. Determine the value of the voltage across AB when the 10 Ω resistor is removed.

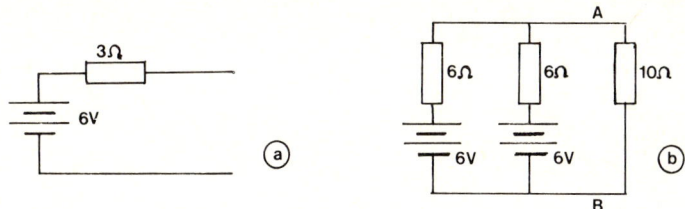

Figure 1.17

Specific objectives

The expected learning outcome is that the student:
1.3 *States maximum power transfer theorem.*
1.4 *States practical sources to which the maximum power transfer theorem is applicable.*

Maximum power transfer theorem

The maximum power transfer theorem states that when one circuit is supplying another, the power transferred from the one circuit to the other is a maximum when the output resistance of the supply circuit is equal to the input resistance of the circuit being supplied.

Consider two circuits shown in block form in fig. 1.18a connected so that circuit A supplies circuit B.

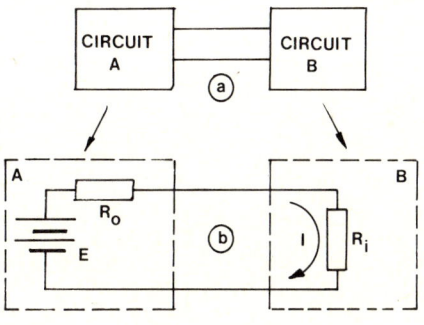

Figure 1.18

Circuit A may be replaced by a Thevenin equivalent consisting of a source E and resistance R_o, circuit B may be replaced by a resistance R_i as shown in fig. 1.18b. The resistance R_o is called the *output resistance* of circuit A, resistance R_i is called the *input resistance* of circuit B.

The current supplied by A to B, shown as I in the figure, is given by

$$I = \frac{E}{R_i + R_o}$$

The power delivered to B, that is, the power in resistance R_i, is given by $I^2 R_i$. Now

$$I^2 R_i = \left(\frac{E}{R_i + R_o}\right)^2 R_i$$

$$= \frac{E^2 R_i}{R_i^2 + 2R_i R_o + R_o^2}$$

The denominator of the right hand side of this equation

$$R_i^2 + 2R_i R_o + R_o^2$$

may be rewritten as

or

$$R_i^2 - 2R_i R_o + R_o^2 + 4R_i R_o$$

$$(R_i - R_o)^2 + 4R_i R_o$$

and so

$$\text{power in } R_i = \frac{E^2 R_i}{R_i^2 + 2R_i R_o + R_o^2}$$

$$= \frac{E^2 R_i}{(R_i - R_o)^2 + 4R_i R_o}$$

For the power to be a maximum, the right hand side of this equation must be a maximum. For this to be so the denominator of the right hand side must be a *minimum*. This occurs when $(R_i - R_o)^2$ is zero since at all other values of $(R_i - R_o)^2$ the denominator has a higher value than $4R_i R_o$, the value it has when $(R_i - R_o)^2$ is zero.

When $(R_i - R_o)^2$ is zero, $(R_i - R_o)$ is zero and $R_i = R_o$, that is, input resistance and output resistance are equàl.

When $R_i = R_o$

$$\text{Maximum power transferred} = \frac{E^2 R_i}{4R_i R_o} = \frac{E^2}{4R_o}$$

Note that the theorem has been demonstrated using d.c. sources and resistances; it is, however, equally true for a.c. sources and when signals are being transferred from one circuit to another within an electronic system.

Example 1.4   Calculate the value of the load resistance to which maximum power may be transferred from the circuit shown in fig. 1.19. Determine also the value of the maximum power.

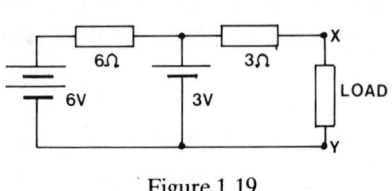

Figure 1.19

When the load is removed, no current flows in the right hand 3 Ω resistor and the voltage across XY is 3 V, the value of the e.m.f. of the right hand battery. (Calculation also shows this: the mesh e.m.f. is $(6 - 3)$ V, the mesh current is

$$\frac{6 - 3}{6} \text{ A i.e. } 0.5 \text{ A}$$

the p.d. across the mesh 6 Ω resistor is 6×0.5 V, i.e. 3 V, so that the voltage across XY, bearing in mind that no current flows in the right hand 3 Ω resistor is the 6 V battery e.m.f. less the 3 V drop across the mesh resistor, i.e. $(6 - 3)$ V, which equals 3 V, the e.m.f. of the right hand battery).

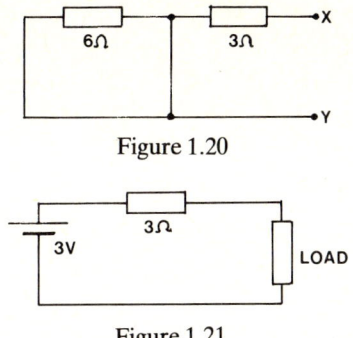

Figure 1.20

Figure 1.21

The Thevenin source voltage is thus 3 V. The resistance of the Thevenin circuit obtained by looking back into the circuit at XY when the sources are replaced by their internal resistances (assumed zero) is 3 Ω, the value of the right hand resistor. See fig. 1.20. Note that the mesh to the left of the 3 V battery is short circuited by the internal resistance of the 3 V battery, assumed zero,

Thus the whole circuit of fig. 1.19 may be replaced by the equivalent circuit of fig. 1.21 and maximum power is transferred when the load resistor has the value 3 Ω, the value of the output resistance of the circuit supplying the load.

$$\text{Maximum power transferred} = \frac{3^2 \times 3}{4 \times 3 \times 3}$$

$$\text{being } \frac{\text{e.m.f.}^2 \times \text{input resistance}}{4 \times \text{input resistance} \times \text{output resistance}} = 0.75 \text{ W}$$

The maximum power transfer theorem is used extensively in audio circuits in which the input resistance of, say, a speaker is made to equal the output resistance of a power amplifier using the impedance changing property of a transformer. The process is called *impedance matching*.

Specific objectives

The expected learning outcome is that the student:
1.5   States Norton's theorem.
1.6   Solves simple problems using Norton's theorem.
1.7   Explains the relationship between Thevenin's theorem and Norton's theorem.

Norton's theorem

Norton's theorem states that any circuit, however complex, may be theoretically replaced by a current generator in parallel with an impedance. The current output of the generator is equal to the current which would flow through a short circuit placed across the output terminals of the circuit to be replaced. The impedance in parallel with the generator has a value equal to the impedance looking back into the output terminals of the circuit to be replaced when all voltage sources in the circuit are replaced by their internal impedances.

As with Thevenin's theorem, Norton's theorem applies equally to all circuits, a.c. or d.c. At frequencies where reactive components of impedances are negligible we may use the word 'resistance' in place of 'impedance' in the theorem.

Norton's theorem is the *converse* of Thevenin's theorem in that the Norton equivalent circuit uses a *current* generator instead of a voltage generator and the impedance (which is the output impedance of the circuit being replaced just as in Thevenin's theorem) is in *parallel* with the generator instead of being in series with it. The only slight difficulty perhaps is an understanding of what is meant by a 'current generator' since we are probably more familiar with voltage generators.

Simply, a current generator is a source generating substantially constant current regardless of load. In practical terms a field effect transistor biased beyond the pinch off point on the drain-current/drain-voltage characteristics acts as a current generator (the characteristic is almost parallel to the voltage axis beyond this point). It must of course be remembered that we are dealing with *theoretical* equivalent circuits when we use network theorems and we cannot always find an *exact* practical equivalent to the generators or impedances in such circuits.

Fig. 1.22 and 1.23 shows Norton's theorem applied to a circuit, the circuit shown in fig. 1.23 being the Norton equivalent of the complex circuit shown in fig. 1.22.

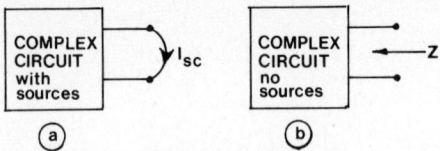

Figure 1.22

In fig. 1.22a I_{sc} is the current which flows when a short circuit is applied across the output of the complex circuit, all sources within the circuit remaining connected. The output impedance, shown as Z in part b of the figure, is the impedance looking back into the complex circuit when all sources are replaced by their internal impedances.

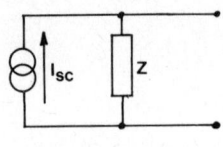

Figure 1.23

Note the symbol used for a current generator in fig. 1.23. When applying Norton's theorem to a circuit containing a number of voltage sources it is usually convenient to replace each voltage source in turn by a current source using the theorem. Let us see how this is done.

Example 1.5   Find the equivalent current sources of the voltage sources shown in fig. 1.24. The internal resistance of each voltage source may be assumed to be 3 Ω.

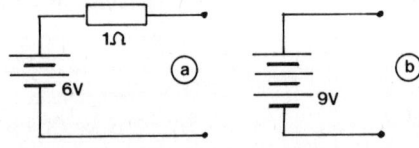

Figure 1.24

If a short circuit is placed across the output terminals of the source shown in fig. 1.24a the current flowing is that due to 6 V applied across 4 Ω, the total resistance (1 Ω + battery internal resistance 3 Ω), i.e. 6/4 or 1.5 A as shown in fig. 1.25a.

Replacing the 6 V source by its internal resistance (3 Ω) and looking back into the circuit, as shown in fig. 1.25b, the output resistance is 4 Ω and the Norton equivalent circuit is as shown in fig. 1.25c.

Similarly, the short circuit current of the source shown in fig.

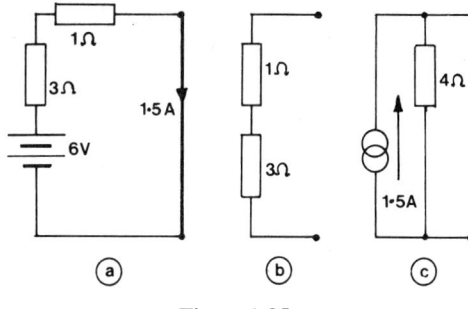

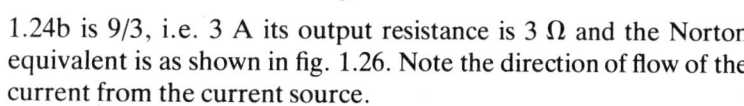

Figure 1.25

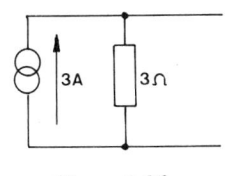

Figure 1.26

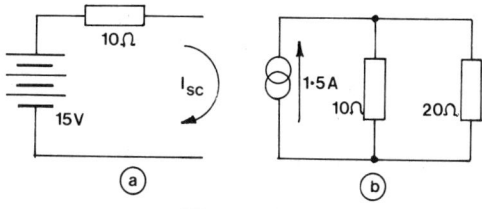

Figure 1.27

1.24b is 9/3, i.e. 3 A its output resistance is 3 Ω and the Norton equivalent is as shown in fig. 1.26. Note the direction of flow of the current from the current source.

Example 1.6 Calculate the current flowing in the 20 Ω load resistor in the circuit shown in fig. 1.27. Check the answer using Norton's theorem.

$$\text{Total circuit resistance} = 10 + 20$$
$$= 30 \ \Omega$$

$$\text{Current flowing in the circuit} = \frac{15}{30}$$
$$= 0.5 \ \text{A}$$

This is the current flowing in the 20 Ω load resistor.

To check this using Norton's theorem we may consider the load to be supplied by a voltage source of 15 V in series with a resistance of 10 Ω and replace this voltage source by a current source as shown in fig. 1.28.

Figure 1.28

The short circuit current is 15/10 i.e. 1.5 A and the output resistance is 10 Ω. The equivalent circuit is thus a 1.5 A current generator and parallel resistance 10 Ω. This supplies the 20 Ω load as shown in fig. 1.28b.

To find the current in the 20 Ω resistor we need to know the p.d. across it. The whole of 1.5 A flows into the parallel combination of 10 Ω and 20 Ω and thus this p.d. = 1.5 × resistance of 10 Ω and 20 Ω in parallel

$$= 1.5 \times \frac{10 \times 20}{10 + 20}$$

$$= \frac{300}{30}$$

$$= 10 \ \text{V}$$

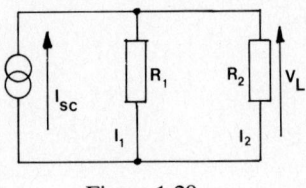

Figure 1.29

The current in the 20 Ω resistor is thus 10/20, i.e. 0.5 A as before.

We would not normally of course use Norton's theorem on such a simple circuit; the technique is useful however when the circuits become more complicated as in the following three examples. These examples are the same as those given earlier in which Thevenin's theorem was used. We shall thus be able to compare the use of these theorems.

Just before we look at the examples it would be useful to revise current distribution in a parallel circuit, since use of Norton's theorem usually gives a final simplified circuit of this form.

Figure 1.29 shows a current generator providing a current I_{sc}, the load on the generator consists of two resistors R_i and R_2 taking current I_1 and I_2, respectively, so that

$$I_{sc} = I_1 + I_2$$

To obtain the p.d. across the load, V_L, multiply total current by the effective load resistance (which is R_1 and R_2 in parallel).

$$V_L = I_{sc} \; \frac{R_1 R_2}{R_1 + R_2}$$

also $\qquad\qquad V_L = I_1 R_1$ and $V_L = I_2 R_2$

To find I_1 divide V_L by R_1

$$I_1 = I_{sc} \; \frac{R_1 R_2}{R_1 + R_2} \; \frac{1}{R_1}$$

$$I_1 = I_{sc} \left(\frac{R_2}{R_1 + R_2} \right)$$

Similarly $\qquad\qquad I_2 = I_{sc} \; \frac{R_1 R_2}{R_1 + R_2} \; \frac{1}{R_2}$

giving $\qquad\qquad I_2 = I_{sc} \left(\frac{R_1}{R_1 + R_2} \right)$

or in words:

Current in one of the two parallel resistors =

$$\frac{\text{total current} \times \text{resistance of other resistor}}{\text{sum of resistance}}$$

This is useful since it avoids having to calculate the load p.d.

Example 1.7   Calculate the current flowing in the 6 Ω resistor in the circuit shown in fig. 1.30 (the same circuit as in fig. 1.3).

Unless given information to the contrary we assume that the resistance in each branch of the circuit containing a voltage source either *is* the source internal resistance or *contains* the source internal resistance.

We may consider the circuit to consist of a 6 Ω load supplied by two sources as shown in fig. 1.31 and we may replace each of these

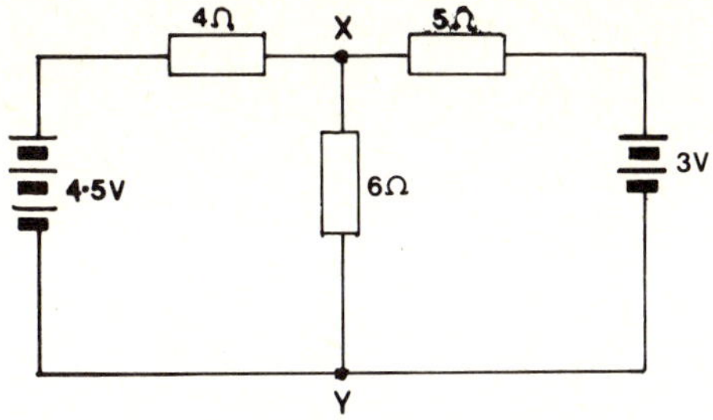

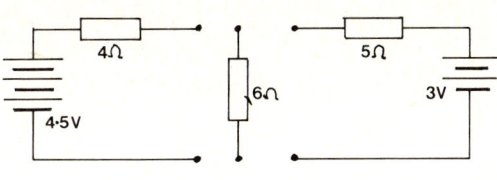

Figure 1.30

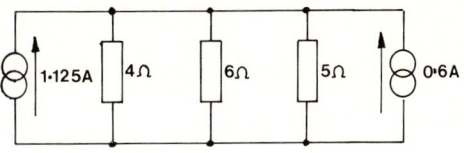

Figure 1.31

voltage sources by an equivalent current source using Norton's theorem. To find the source current, short circuit the output in each case which gives us

$$\frac{4.5}{4} \text{ i.e. } 1.125 \text{ A}$$

for the 4.5 V source and

$$\frac{3}{5} \text{ i.e. } 0.6 \text{ A}$$

for the 3 V source. The output resistances of the current sources are 4 Ω and 5 Ω, respectively, so that we may redraw fig. 1.30 as in fig. 1.32.

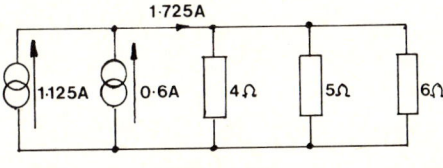

Figure 1.32

The circuit can be seen to consist of a load made up of the parallel combination of three resistors supplied by two sources, one giving 1.125 A, the other giving 0.6 A. The sum of these currents flows into the load as shown in fig. 1.33.

Figure 1.33

We may replace the current sources by a single source providing (1.125 + 0.6), i.e. 1.725 A and we may replace the 4 Ω/5 Ω parallel combination by a single resistor of value

$$\frac{4 \times 5}{4 + 5} \text{ i.e. } 2.22 \ \Omega$$

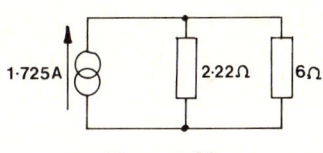

Figure 1.34

giving the simplified circuit of fig. 1.34.

The current in the 6 Ω resistor (using the formula derived earlier) is thus given by

$$\frac{2.22}{2.22 + 6} \times 1.725$$

which gives 0.466 A as before.

Example 1.8   Calculate the p.d. across XY in the circuit of fig. 1.35 (this is the same problem as Example 1.2).

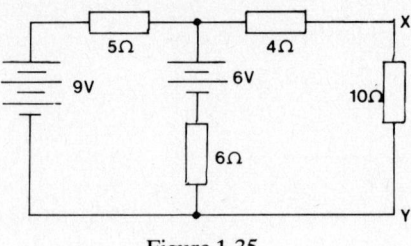

Figure 1.35

We may replace the 9 V, 5 Ω voltage source by a 9/5 A, 5 Ω current source and the 6 V, 6 Ω source by a 6/6 A, 6 Ω current source as shown in fig. 1.36.

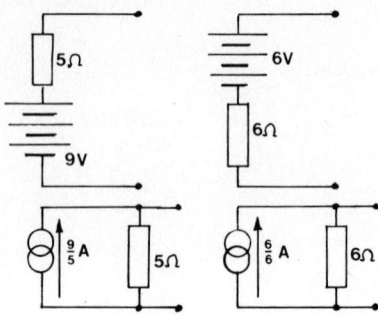

Figure 1.36

and redraw fig. 1.35 as in fig. 1.37

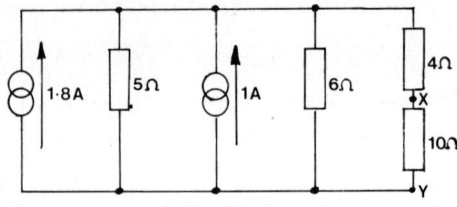

Figure 1.37

The two current generators may be replaced by a single generator providing 2.8 A (the two generators replaced act in the same direction) and a parallel resistance equal to the parallel combination of 5 Ω and 6 Ω, i.e.

$$\frac{5 \times 6}{5 + 6}$$

which equals 2.73 Ω.

See fig. 1.38.

The resistance of the branch containing XY is 4 + 10, i.e. 14 Ω and the current in branch XY is given by

$$\frac{2.73}{2.73 + 14} \times 2.8$$

which equals 0.457 A.

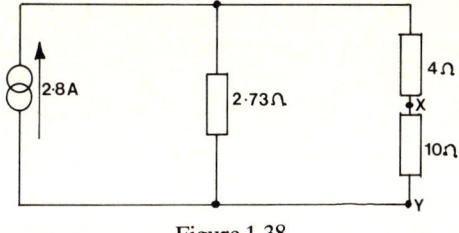

Figure 1.38

The p.d. across XY is the product of this current and the resistance of XY (10 Ω).

$$\text{p.d. across XY} = 0.457 \times 10$$
$$= 4.57\,\text{V}$$

Example 1.9 Find the current in the 20 Ω resistor in the circuit shown in fig. 1.39 (this is the same problem as Example 1.3).

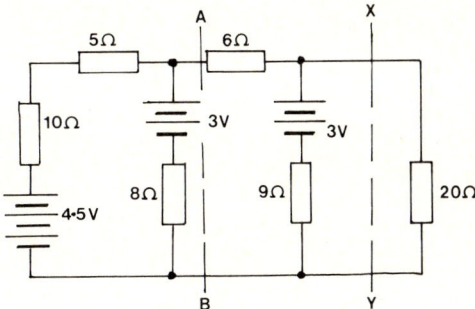

Figure 1.39

The same technique is used as in the previous example. In this one, however, extra care must be taken to make sure the connections in the equivalent circuit are exactly as in the circuit being replaced. There is a 6 Ω resistor between the two 3 V sources which cannot be included as part of the replacement current generator of either of them because of the connection of the 4.5 V, 15 Ω source on the left hand side of the figure and the connection of the 20 Ω load on the right hand side of the figure.

The three voltage sources in the circuit may be replaced by current sources as shown in fig. 1.40.

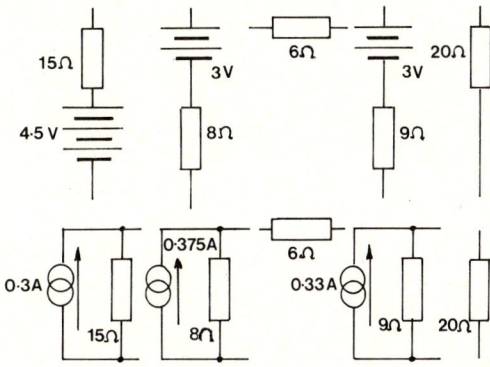

Figure 1.40

Note that the 10 Ω and 5 Ω resistors have been replaced by a 15 Ω resistor, which forms the output resistance of the 4.5 V source. The currents of the current sources, namely, 0.3 A, 0.375 A and 0.33 A are obtained by short circuiting each voltage source and its resistance in turn to obtain

$$\frac{4.5}{1.5} \text{A}, \frac{3}{8} \text{A and } \frac{3}{9} \text{A}$$

respectively.

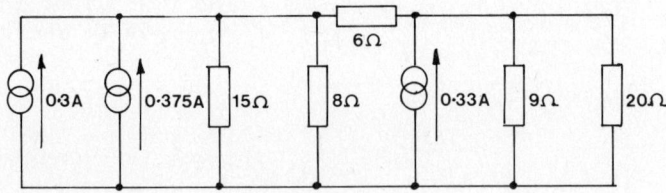

Figure 1.41

The circuit may be redrawn as shown in fig. 1.41 and the two current generators at the left hand side of the 6 Ω resistor may be replaced by a single current generator providing (0.3 + 0.375) A, i.e. 0.675 A and having a single parallel resistance equal to the parallel combination of 15 Ω and 8 Ω, i.e.

$$\frac{8 \times 15}{8 + 15}$$

which equals 5.22 Ω. See fig. 1.42.

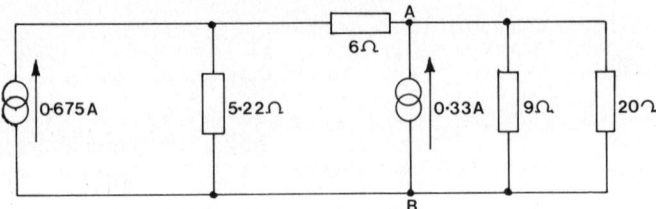

Figure 1.42

We now apply Norton's theorem again to the left of AB in fig. 1.42 to obtain a single current generator in parallel with a single resistance, to replace the two resistors 5.22 Ω and 6 Ω.

To do this, first short circuit AB to obtain the output current of the new single generator.

The current in the short circuit

$$I = \frac{5.22}{6 + 5.22} \times 0.675 \text{ A}$$

since the circuit consists of a 0.675 A generator supplying a load of 5.22 Ω in parallel with 6 Ω. Thus

$$I = 0.31 \text{ A}$$

The output resistance of the 0.675 A generator is 5.22 Ω; thus, the output resistance of the replacement generator may be obtained by replacing the 0.675 A generator by 5.22 Ω and looking back into

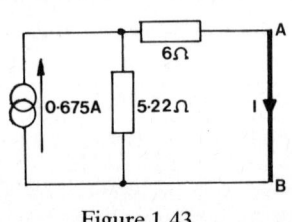

Figure 1.43

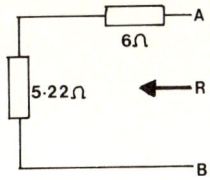

Figure 1.44

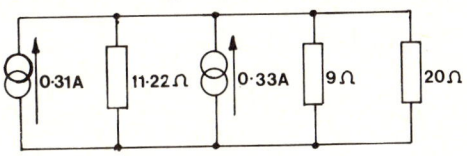

Figure 1.45

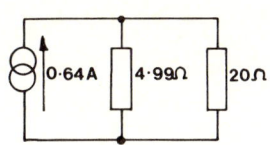

Figure 1.46

the circuit to the left of AB as shown in fig. 1.44 to give $R = (6 + 5.22)\ \Omega$, i.e. $11.22\ \Omega$.

We now have a circuit consisting of a 0.31 A, 11.22 Ω current source in parallel with a 0.33 A, 9 Ω current source feeding the 20 Ω load as shown in Fig. 1.45 and, taking into account the direction of the current supplied, we may replace the two current sources by a single current source of $(0.31 + 0.33)$ A, i.e. 0.64 A, the parallel resistance of the new source being equal to the parallel combination of 11.22 Ω and 9 Ω,

i.e. $$\frac{11.22 \times 9}{11.22 + 9}\ \Omega$$

which equals 4.99 Ω

The final simplified circuit is thus as shown in fig. 1.46 and the required current in the 20 Ω load is

$$\frac{4.99}{4.99 + 20} \times 0.64$$

which equals 0.129 A as before.

This example is rather more complicated than the previous one in that successive applications of Norton's theorem were required. Provided that care is taken always to maintain correct connections in the equivalent circuits as they are obtained, the process may be continued indefinitely for any circuit, however complex, the ultimate aim always being to obtain a single generator connected to the resistance the current in which or the p.d. across which has to be found.

Summary

Thevenin's theorem

Any circuit, however complex, may be replaced by a voltage generator in series with an impedance. The generated e.m.f. is the voltage which appears across the output terminals of the circuit to be replaced when any load is removed. The impedance in series with the generator has a value equal to the impedance looking back into the output terminals of the circuit to be replaced when all voltage sources in the circuit are replaced by their internal impedances.

Maximum power transfer theorem

The maximum power transfer theorem states that when one circuit is supplying another the power transferred from the one circuit to the other is a maximum when the output resistance of the supply circuit is equal to the input resistance of the circuit being supplied.

Norton's theorem

Any circuit, however complex, may be replaced by a current generator in parallel with an impedance. The current output of the generator is equal to the current which would flow through a short circuit placed across the output terminals of the circuit to be replaced. The impedance in parallel with the generator has a value equal to the impedance looking back into the output terminals of the circuit to be replaced when all voltage sources in the circuit are replaced by their internal impedances. Norton's theorem is the converse of Thevenin's theorem.

EXERCISE 1.2

1. Using Norton's theorem calculate the current in the 5 Ω resistor of the circuit shown in fig. 1.13.

2. Calculate the voltage across XY in the circuit shown in fig. 1.14 using Norton's theorem.

3. Determine the Norton equivalent circuit of the circuit shown in fig. 1.15.

4. The Norton equivalent circuit of the circuit shown in fig. 1.47 has a parallel resistance of value 10 Ω. What is the value of *R*?

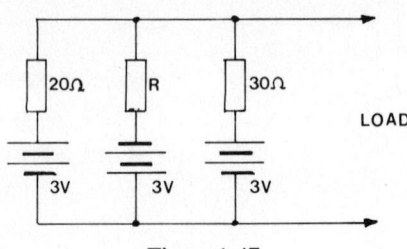

Figure 1.47

5. The circuit shown in fig. 1.48a is the Norton equivalent of the circuit to the left of AB in fig. 1.48b. Determine the current which would flow in a short circuit placed across AB.

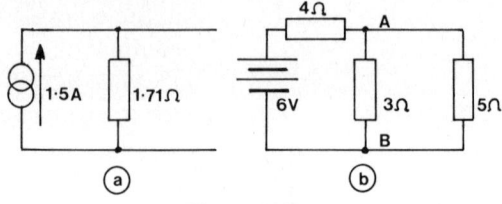

Figure 1.48

Multiple-choice examples

Example 1.10   The output resistance of the circuit shown in fig. 1.49 if the battery internal resistance is zero is

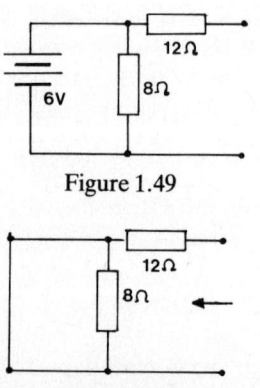

Figure 1.49

A. 8 Ω   C. 20 Ω
B. 12 Ω   D. 4.8 Ω

To obtain the output resistance we replace the battery by its internal resistance, i.e. 0 Ω, and look back into the circuit as in fig. 1.50.

A. The 8 Ω resistor is short circuited, this answer is therefore incorrect.

B. Since the 8 Ω resistor is short circuited the resistance looking back into the circuit is 12 Ω in series with 0 Ω, i.e. 12 Ω. This answer is therefore correct.

Figure 1.50

C. This answer is derived by finding the equivalent of 12 Ω and 8 Ω in series; the 8 Ω resistor has been replaced by a short circuit, however, so the method and the answer are incorrect.

D. This answer is derived by finding the equivalent of 12 Ω and 8 Ω in parallel; the 8 Ω resistor has been replaced by a short circuit, however, so the method and the answer are incorrect.

Example 1.11   If the maximum power were to be transferred to a load by the circuit of fig. 1.49, what would be its value?

A. 8 Ω   C. 20 Ω

B. 12 Ω   D. 4.8 Ω

This problem is another way of saying 'what is the output resistance of the circuit?' since maximum power is transferred to a load when its value is equal to the circuit output resistance. The comments on each suggested answer given in Example 1.10 still apply. Answer B is correct.

Example 1.12   A 4 V, 3 Ω voltage source is connected in parallel with a 6 Ω resistor. The parallel resistance of the Norton equivalent circuit is

A. 9 Ω   C. 3 Ω

B. 6 Ω   D. 2 Ω

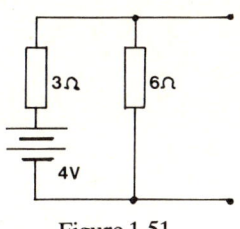

Figure 1.51

The parallel resistance of the Norton equivalent circuit is obtained by replacing the source in the original circuit by its internal resistance, 3 Ω, and finding the equivalent resistance of the circuit as a whole. When the source is replaced by 3 Ω we have 3 Ω in parallel with 6 Ω. Answer D is therefore correct, the remaining answers being incorrect.

Example 1.13   The Norton equivalent of a certain circuit consists of a 5 A current generator in parallel with a 4 Ω resistor. What is the Thevenin equivalent of this circuit?

A. A 20 V source in series with a 4 Ω resistor.

B. A 0.8 V source in series with a 4 Ω resistor.

C. A 1.25 V source in series with a 4 Ω resistor.

D. It cannot be determined without further information.

A. To replace a voltage source in series with a resistor (the Thevenin equivalent circuit) by a Norton equivalent we short circuit the output to determine the current of the Norton current generator. The current is thus the figure obtained by dividing the voltage of the voltage source by the resistance of the series resistor. The Norton resistance has the same value as the Thevenin resistance although it is connected differently (parallel connection not series). Thus, here we have a 5 A, 4 Ω current generator so that the voltage which would cause 5 A to flow in 4 Ω is 20 V (5 × 4) and this is the e.m.f. of the Thevenin voltage source. Answer A is therefore correct.

B and C. These answers are derived by the misapplication of Ohm's law. In B the resistance has been divided by current and in C the current has been divided by resistance to give the figure indicated as the Thevenin voltage. Correctly, of course, the current and resistance should be multiplied together.

D. This statement is untrue. Knowing the content of one equivalent circuit we are always able to find the other.

Example 1.14 If two identical 2 A, 3 Ω current sources are connected in parallel the single equivalent current source is

A. 2 A, 3 Ω C. 4 A, 1.5 Ω

B. 4 A, 6 Ω D. 4 A, 3 Ω

If the sources are identical the current supplied by the single replacement generator is the *sum* of the individual currents of the generators being replaced; the output resistance of the single replacement source is the parallel combination of the resistances of the sources being replaced.

A. Both current and resistance are incorrect. Parallel connection of identical current sources must produce a changed current and resistance in the single equivalent circuit.

B. The current is correct, the individual resistances have been added however as if they acted in series. This is incorrect.

C. The combined current is 2 + 2, i.e. 4 A. The parallel combination of two 3 Ω resistors yields a resistance of $(3 \times 3)/(3 + 3)$ i.e. 1.5 Ω. This answer is therefore correct.

D. The current is correct, the resistance incorrect.

EXERCISE 1.3

1. A source consisting of two batteries in parallel, like polarity to like, one 5 V, 1 Ω, the other 3 V, 0.5 Ω supplies current to a 15 Ω load. Calculate the value of the current using Thevenin's theorem.

2. A 10 V, 2 Ω battery is connected across a 12 Ω load. A 0–10 Ω continuously variable resistor is connected across the load. Calculate the resistance of the variable resistor when the source is delivering maximum power.

3. A voltage source providing E volts and having an internal resistance of 20 Ω is connected in series with a resistor of resistance R ohms. When a 50 Ω resistor is connected across this resistor maximum power of 8 W is obtained from the source. Calculate the values of E and R.

4. The open circuit e.m.f. of a battery is 15 V. When connected to a 30 Ω load the battery voltage falls to 14 V. Calculate the load current when four of these batteries are connected in parallel to supply a 10 Ω load.

5. The current generator equivalent circuit of a certain battery consists of a 5 A source in parallel with a 3 Ω resistor. Use Thevenin's theorem to calculate the load current when three of these batteries are connected in parallel to supply a 10 Ω load.

6. Calculate the current in the 20 Ω resistor in the circuit shown in fig. 1.52.

7. Calculate the maximum power delivered to the variable resistor in the circuit shown in fig. 1.53 as it is varied throughout its range and the value of its resistance when maximum power is supplied to it.

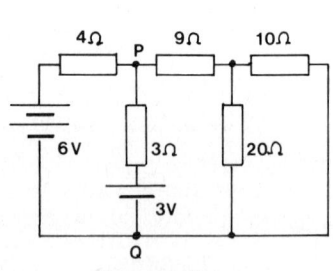

Figure 1.52

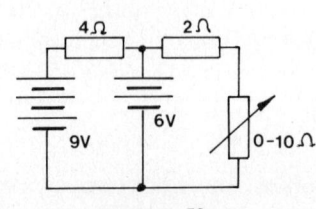

Figure 1.53

8. Two 10 V, 2 Ω batteries are connected in series aiding. The single equivalent current generator to replace this combination is

    A. 5 A, 2 Ω   C. 5 A, 1 Ω
    B. 5 A, 4 Ω   D. 10 A, 4 Ω

9. The maximum power delivered by a certain 8 V battery is 16 W. The battery internal resistance is

    A. 2 Ω   C. 4 Ω
    B. 1 Ω   D. incalculable without further information.

10. Two identical 4 V, 1 Ω batteries are connected in parallel. The current and output resistance of the single replacement current generator respectively are

    A. 4 A, 1 Ω   C. 4 A, 0.5 Ω
    B. 8 A, 0.5 Ω   D. 8 A, 1 Ω

Possible marks

SELF-ASSESSMENT EXERCISE 1

1. State Thevenin's theorem. (3)

2. Two 4.5 V batteries each of internal resistance 2 Ω are connected in series. What is the Thevenin equivalent circuit of the combination? (3)

3. State Norton's theorem. (3)

4. Two 4.5 V batteries each of internal resistance 2 Ω are connected in parallel, like polarity to like. What is the Norton equivalent circuit of the combination? (3)

5. State the maximum power transfer theorem. (3)

6. A 4.5 V, 2 Ω battery is connected in series with a 3 V, 1 Ω battery. What is the value of resistance to which the combination will deliver maximum power? Calculate the maximum power. (5)

7. Two identical batteries connected in series deliver a maximum power of 1 W to a resistor of resistance 4 Ω. What is the open circuit e.m.f. and internal resistance of each battery? (5)

8. The Norton equivalent circuit of two identical batteries connected in parallel consists of a 1 A current generator in parallel with a 4 Ω resistor. Calculate the value of the resistive load to which a single battery will deliver maximum power and the value of the maximum power. (5)

9. Use Thevenin's theorem to find the current flowing in the 20 Ω resistor in the circuit of fig. 1.8 when the two 3 V batteries have their connections reversed. (14)

10. Use Norton's theorem to calculate the current flowing in the 5 Ω resistor in the circuit of fig. 1.14 when the right hand 3 V battery has its connections reversed. (14)

11. If in the circuit of fig. 1.19 the 6 Ω and 3 Ω resistances are interchanged, calculate the maximum power delivered from the combination and the value of the load resistance when this occurs. (14)

12. Calculate, using Norton's theorem, the current flowing in the 3 Ω resistor in the circuit of fig. 1.52. (14)

13. The Thevenin equivalent of a certain circuit consists of a 5 V d.c. source in series with a resistance of 2 Ω. The Norton equivalent of a second circuit is a 1.5 A current generator in parallel with a resistance of 0.5 Ω. The two currents are connected in parallel like polarity to like.

Determine:
(a) the Norton equivalent of the combination;
(b) the Thevenin equivalent of the combination;
(c) the maximum power which can be delivered by the combination and the value of the load resistance when the power delivered is a maximum.

(14)

Answers

EXERCISE 1.1
1. 0.087 A
2. 1.44 V
3. I V source; series resistance 13.33 Ω
4. 60 Ω
5. 6 V

EXERCISE 1.2
1. 0.087 A
2. 1.44 V
3. 0.225 A source, parallel resistance 13.33 Ω
4. 60 Ω
5. 1.5 A

EXERCISE 1.3
1. 0.24 A
2. 2.4 Ω
3. 25.29 V; 33.33 Ω
4. 1.42 A
5. 1.36 A
6. 0.082 A
7. 4.5 W; 2 Ω
8. B
9. B
10. B

Marks

SELF-ASSESSMENT EXERCISE 1

1. The statement should include the theorem (1), how the value of the generator e.m.f. is determined (1) and how the value of the series impedance is determined (1). (3)

2. A 9 V battery (1½) in series with a 4 Ω resistance (1½). (3)

3. The statement should include the theorem (1), how the value of the generator current is determined (1) and how the value of the parallel impedance is determined (1). (3)

4. A 4.5 A current generator (1½) in parallel with a 1 Ω resistance (1½). (3)

The 4.5 A is determined by adding together the current through a short circuit placed across the output terminals of the circuit due to one battery (4.5/2 A) to the current through the short circuit due to the other battery (4.5/2 A), i.e. 2.25 + 2.25 A. The parallel resistance is made up of 2 Ω in parallel with 2 Ω.

5. Statement as given in the text. (3)

6. The equivalent d.c. source has an open circuit e.m.f. of 4.5 V + 3 V, i.e. 7.5 V and a series resistance of 2 + 1, i.e. 3 Ω. (1) (1)
Maximum power is delivered when the load and the source series resistance are the same, i.e. when the load has the value 3 Ω. (1)
The value of the maximum power is

$$\left(\frac{7.5}{6}\right)^2 \times 3, \text{ i.e. } 4.69 \text{ W} \tag{2}$$

7. The series resistance of the single equivalent source is 4 Ω since this is the value of load to which maximum power is delivered. The series resistance of each battery is therefore 2 Ω since there are two identical batteries. (1) (1)
The maximum power delivered is

$$\left(\frac{E}{8}\right)^2 \times 4, \text{ i.e. } E^2/16$$

where E is the open circuit e.m.f. of the single equivalent source, and this is equal to 1 W. Therefore

$$\frac{E^2}{16} = 1 \text{ and } E = 4 \text{ V} \tag{2}$$

The single equivalent source replaces two identical batteries connected in series; the open circuit e.m.f. of each battery is therefore 4/2, i.e. 2 V. (1)

8. Each battery will supply ½ × 1 A short circuit current, i.e. ½ A. The parallel combination of the two single battery internal resistances is 8 Ω so that each single battery internal resistance is 8 Ω (two 8 Ω resistances in parallel combine to give 4 Ω). The single battery open circuit e.m.f. must be then that e.m.f. which provides ½ A through 8 Ω, i.e. 4 V. Each single battery is therefore of 4 V e.m.f., internal resistance 8 Ω. Value of resistance to which a single battery will deliver maximum power is therefore 8 Ω and the maximum power delivered is

$$\left(\frac{4}{16}\right)^2 \times 8, \text{ i.e. } 0.5 \text{ W} \qquad (1) (1) (1) (2)$$

9. Current I_1 in fig. 1.10 with 3 V battery reversed is given by

$$I_1 = \frac{7.5}{23} \text{ i.e. } 0.326 \text{ A} \tag{2}$$

Voltage across AB is therefore

$$4.5 - (0.326 \times 15) \text{ i.e. } -0.39 \text{ V} \tag{2}$$

The equivalent resistance is the parallel combination of 15 Ω and 8 Ω, i.e. 5.22 Ω, as before. (2)

In the circuit of fig. 1.11, the 3.52 V battery is now replaced by a battery of e.m.f. −0.39 V (i.e. 0.39 V with the positive pole connected to point B). The 3 V battery must also be reversed as given in the question.

Current I_2 in the revised fig. 1.11 (with the 20 Ω load removed) is now

$$\frac{(3 - 0.39)}{5.22 + 6 + 9} = 0.129 \text{ A} \tag{2}$$

and the voltage across XY is 3 − (9 × 0.129), i.e. 1.84 V. (2)

The equivalent resistance is (5.22 + 6) in parallel with 9, i.e. 4.99 Ω as before. (2)

The equivalent circuit is now as fig. 1.12 with the 3.23 V source replaced by a 1.84 V source and the current in the 20 Ω load is 1.84/24.99 i.e. 0.0736 A. (2)

10. The left hand 3 V battery in fig. 1.14 may be replaced by a current generator providing 3/6, i.e. 1/2 A in parallel with a 6 Ω resistance.   (2)
The right hand 3 V battery (which has been reversed) may similarly be replaced.   (2)
The 4.5 V battery may be replaced by a current generator providing 4.5/4, i.e. 1.125 A in parallel with a 4 Ω resistance.   (2)

The Norton equivalent circuit now becomes as fig. 1.54 (note the direction of the generator currents)

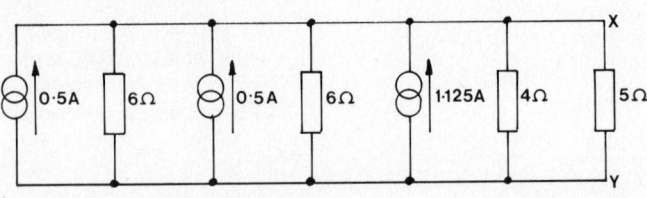

Figure 1.54

the single Norton equivalent generator being one providing 0.5 + 0.5 + 1.125, i.e. 2.125 A with a parallel resistance equal to 6 Ω, 6 Ω and 4 Ω in parallel, i.e. 1.71 Ω.   (2) (2)

From the circuit of fig. 1.55 current in 5 Ω resistor is

$$\frac{1.71}{1.71 + 5} \times 2.125 \text{ i.e. } 0.542 \text{ A} \tag{2}$$

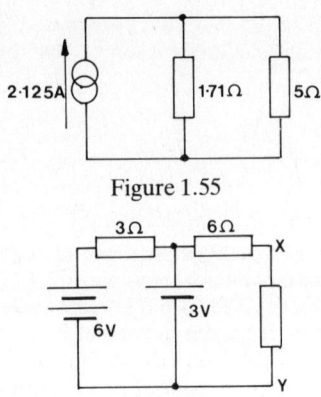

Figure 1.55

Figure 1.56

11. The circuit is as fig. 1.56. When the load is removed no current flows in the 6 Ω resistor. The open circuit e.m.f. across XY is thus 3 V (the e.m.f. of the right hand battery).   (2) (2)

Replacing the batteries by their internal resistances (assumed zero) the 3 Ω resistor is short circuited and the resistance looking back into XY is therefore 6 Ω.   (2) (2)

The Thevenin equivalent source is therefore 3 V with a series resistance of 6 Ω.   (2)

Maximum power is delivered when the load is 6 Ω, the value of the maximum power being

$$\left(\frac{3}{12}\right)^2 \times 6 \text{ i.e. } 0.375 \text{ W} \tag{2}$$

12. The circuit may be considered to be made up of a 6 V, 4 Ω source in parallel with a 3 V, 3 Ω source in parallel with the network composed of the 9 Ω, 10 Ω and 20 Ω resistors.   (2)

The network resistance is 9 Ω in series with the parallel combination of 10 Ω and 20 Ω, i.e. 15.67 Ω. See fig. 1.57.   (2)

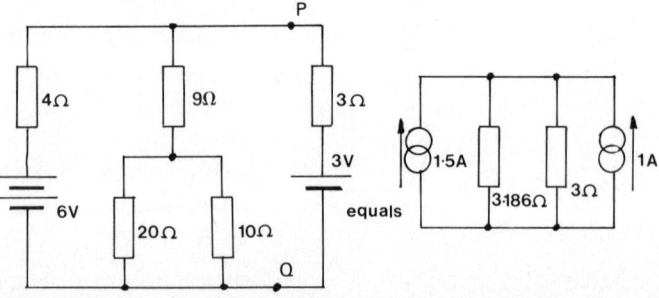

Figure 1.57

Consider the 6 V, 4 Ω source in parallel with the network (equivalent resistance 15.67 Ω) alone. Short circuiting PQ the short circuit current is 6/4, i.e. 1.5 A. The resistance looking into PQ (the 3 V, 3 Ω source removed and the 6 V source short-circuited) is 4 Ω in parallel with 15.67 Ω, i.e. 3.186 Ω. The Norton equivalent is thus a 1.5 A generator in parallel with a 3.186 Ω resistance. (2) (2)

The Norton equivalent of the 3 V, 3 Ω source is a 1 A current generator in parallel with a 3 Ω resistance. From fig. 1.57 the current in the 3 Ω resistance is thus

$$\frac{3.186}{3 + 3.186} \times (1.5 + 1) \text{ i.e. } 1.287 \text{ A} \qquad (2)\,(2)\,(2)$$

13. (a) The Thevenin circuit consisting of 5 V, 2 Ω source may be replaced by a Norton current generator 2.5 A, 2 Ω. This is placed in parallel with a 1.5 A, 0.5 Ω current generator. The overall Norton equivalent is then a current generator providing 2.5 A + 1.5 A, i.e. 4 A and having a parallel resistance of 2 Ω in parallel with 0.5 Ω, i.e. 0.4 Ω. (2) (2)

(b) The 1.5 A, 0.5 Ω current generator may be replaced by a Thevenin voltage source of 0.75 V, 0.5 Ω. This is placed in parallel with a second source of 5 V, 2 Ω like polarity to like. The mesh voltage is (5 − 0.75), i.e. 4.25 V, the mesh current is 4.25/2.5 (the 2.5 being the mesh resistance), i.e. 1.7 A. The output voltage of the combination is thus

$$5 - 2 \times 1.7 \text{ i.e. } 1.6 \text{ V}$$

or

$$0.75 + 0.5 \times 1.7 \text{ i.e. } 1.6 \text{ V} \qquad (2)\,(1)\,(1)\,(2)$$

and the equivalent resistance is 0.5 Ω in parallel with 2 Ω, i.e. 0.4 Ω. (2)

This may be obtained directly by using the answer of part (a). It is not however recommended to use answers already calculated unless it is absolutely necessary (as in Part (c)).

(c) Maximum power is delivered to a 0.4 Ω load its value being

$$\left(\frac{1.6}{0.8}\right)^2 \times 0.4 = 1.6 \text{ W} \qquad (1)\,(1)$$

2 Alternating current circuits

Topic Area: B

General objective
The expected learning outcome is that the student applies a.c. circuit theory to the solution of series and parallel networks.

Specific objectives
The expected learning outcome is that the student:
2.1 *Draws phasor diagrams for series combinations of inductive and capacitive impedances taking account of their resistances.*
2.2 *Determines the current and its phase angle relative to the supply voltage for the circuits in 2.1.*
2.3 *Determines the potential differences across each of the elements in 2.1.*
2.4 *Derives the power triangle from the voltage triangle.*
2.5 *Identifies the active and reactive components of current power.*
2.6 *Defines power factor.*
2.19 *Determines the power dissipated in the circuits of 2.1 and 2.11.*

Series circuits

The theory of series a.c. circuits is contained in detail in *Electrical and Electronic Principles 2* since the specific objectives listed above also form part of that unit. The important points are summarised below.

The impedance of a circuit is the opposition of the circuit to the flow of alternating current. The usual symbol is Z and the quantity is measured in ohms; it is obtained by dividing the voltage applied to the circuit by the current flowing through it. Circuit impedance may contain one or both of two quantities, resistance, symbol R and reactance, symbol X. Resistance is the opposition to current of a resistor, reactance is the opposition to current due to inductance or capacitance. Both are measured in ohms.

In general

$$Z^2 = R^2 + X^2$$

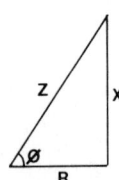

Figure 2.1

the impedance diagram being shown in fig. 2.1. X is the *resultant* reactance, being the difference between the inductive reactance X_L and the capacitive reactance X_C, so that

$$Z^2 = R^2 + (X_L \sim X_C)^2$$

where the $\sim$ sign means 'the difference between'.

Phase angle ϕ is the phase displacement between applied voltage and current. For a circuit containing resistance and reactance

$$\sin \phi = \frac{X}{Z}, \cos \phi = \frac{R}{Z}, \tan \phi = \frac{X}{R}$$

where X is the resultance reactance, i.e. is X_L if the circuit contains inductance but not capacitance, X_C if the circuit contains capacitance but not inductance and $(X_L \sim X_C)$ if the circuit contains both.

Inductive reactance

$$X_L = 2\pi f L$$

and capacitive reactance

$$X_C = 1/2\pi f C$$

where f (Hz) is the supply frequency, L (henrys) is the inductance and C (farads) is the capacitance.

Power in a series a.c. circuit

Power P developed in a circuit is given by

$$V_S I_S \cos \phi \quad (W)$$

where V_S and I_S represent applied voltage and resultant current, ϕ is the phase angle and $\cos \phi$ is called the *power factor*.

The circuit volt-amperes, S, are given by

$$S = V_S I_S \quad (VA)$$

and the reactive volt-amperes, Q are given by

$$Q = V_S I_S \sin \phi \quad (VA_r)$$

The units should be carefully noted in each case. The power triangle is shown in fig. 2.2c from which

$$S^2 = P^2 + Q^2$$

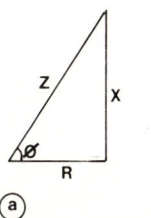
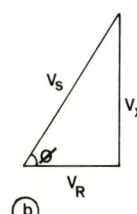
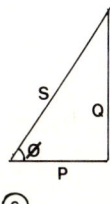

Figure 2.2

This triangle is derived from the impedance triangle shown in fig. 2.1, each side having been multiplied by the square of the circuit current, from which

$$S = Z I_S^2 = Z I_S I_S = V_S I_S$$
$$P = R I_S^2 = R I_S I_S = V_R I_S$$
$$Q = X I_S^2 = X I_S I_S = V_X I_S$$

where V is the total applied voltage, V_R is the component voltage across the resistance and V_X the component voltage across the reactance.

From fig. 2.2b

$$V_R = V_S \cos \phi \text{ and } V_X = V_S \sin \phi$$

Example 2.1   An inductive coil of inductance 10 H and resistance

150 Ω is connected in series with a 5 μF capacitor and a 100 Ω resistor across a 240 V, 50 Hz supply. Calculate the:

(a) inductive reactance   (d) circuit current
(b) capacitive reactance   (e) power factor
(c) impedance   (f) power, volt amperes
                                       and reactive volt amperes.

(a) inductive reactance, $X_L = 2\pi \times 50 \times 10$
$$= 3141.6\,\Omega$$

(b) capacitive reactance, $X_C = 1/2\pi \times 50 \times 5 \times 10^{-6}$
$$= 636.6\,\Omega$$

(c) circuit reactance, $X = X_L - X_C$
$$= 3141.6 - 636.6$$
$$= 2505\,\Omega$$
circuit resistance, $R = 150 + 100$
$$= 250\,\Omega$$
impedance, $Z = \sqrt{(R^2 + X^2)}$
$$= \sqrt{(250^2 + 2505^2)}$$
$$= 2517.4\,\Omega$$

(d) current, $I = V/Z$ (where V is the applied voltage)
$$= 240/2517.4$$
$$= 0.095\text{ A}$$

(e) power factor, $\cos\phi = R/Z$
$$= 250/2517.4$$
$$= 0.099$$

(f) power, $P = VI\cos\phi$
$$= 240 \times 0.095 \times 0.099$$
$$= 2.264\text{ W}$$
volt amperes, $S = VI$
$$= 240 \times 0.095$$
$$= 22.8\text{ VA}$$
reactive volt-amperes, $Q = VI\sin\phi$
$$= 240 \times 0.095 \times 0.995$$
(where $\sin\phi = 0.995$, obtained by tables or calculator from knowing that $\cos\phi = 0.099$)
$$= 22.69\text{ VAr}$$

Specific objectives

The expected learning outcome is that the student:
2.7   *States the conditions for series resonance.*
2.8   *Derives and applies the formula for the series resonance frequency.*
2.9   *Defines Q-factor as the voltage magnification in a series circuit.*
2.10   *Explains:*
       (a) advantages of high Q-factor in series power circuits.
       (b) disadvantages of high Q-factor in series power circuits.

Series resonance The impedance of a series circuit containing resistance, inductance and capacitance

$$Z = \sqrt{[R^2 + (X_L \sim X_C)^2]}$$

When $X_L = X_C$ *series resonance* is said to occur.

At resonance, $X_L = X_C$ so that

$$2\pi f_r L = 1/2\pi f_r C$$

where f_r represents the resonant frequency. Hence

$$f_r^2 = \frac{1}{4\pi^2 LC}$$

and

$$f_r = \frac{1}{2\pi\sqrt{(LC)}}$$

At resonance the voltage across the inductive reactance, V_L, is equal and opposite (in phase) to the voltage across the capacitive reactance, V_C. These voltages cancel so that the supply voltage, V_S, which is given by

$$\overline{V}_S = \overline{V}_R + (\overline{V}_L \sim \overline{V}_C)$$

where V_R is the voltage across the resistance and the bar over each variable symbol indicates a phasor quantity, is equal to V_R, i.e.

$$\overline{V}_S = \overline{V}_R$$

The voltage V_L (which is equal to and opposite to V_C) may be many times larger than V_R *and thus than* V_S. This is because

$$V_L = IX_L \text{ and } V_C = IX_C$$

where I represents circuit current, and

$$V_S = V_R = IR$$

and X_L and X_C, which are dependent on frequency, may be many times larger than R which has a fixed value independent of frequency.

This phenomenon of a series circuit at resonance is called *voltage magnification*.

The voltage magnification, that is, the number of times that V_L or V_C is larger than V_S, is called the *Q-factor* of the circuit and is given the symbol Q.

$$Q = \frac{V_L}{V_S} \text{ or } \frac{V_C}{V_S}$$

but $V_S = V_R$, so that

$$Q = \frac{V_L}{V_R} \text{ or } \frac{V_C}{V_R}$$

$$= \frac{IX_L}{IR} \text{ or } \frac{IX_C}{IR}$$

$$= X_L/R \text{ or } X_C/R$$

and since

$$Q = 2\pi f_r L/R \text{ or } 1/2\pi f_r CR$$

$$f_r = \frac{1}{2\pi\sqrt{(LC)}}$$

$$Q = \frac{2\pi L}{[2\pi\sqrt{(LC)}]R} \text{ or } \frac{2\pi\sqrt{(LC)}}{2\pi CR}$$

$$= \frac{1}{R}\sqrt{(L/C)}$$

The importance of Q as a measure of the frequency sensitive characteristics of a series resonant circuit is considered later in the chapter. There we shall see how use is made of the property in radio-frequency circuits.

Radio-frequency circuits employing the frequency sensitive properties due to series resonance are normally operated at low power. In series circuits used elsewhere at high power, resonance is normally avoided due to the risk to insulation caused by the voltage magnification at resonance, the higher the value of Q the greater being the risk.

Example 2.2   A 5 H, 150 Ω inductor is connected in series with a 0.5 μF capacitor. Calculate the:
  (a) resonant frequency;
  (b) Q-factor of the circuit;
  (c) phase angle when the frequency is twice the resonant frequency.

(a) Resonant frequency $f_r = \dfrac{1}{2\pi\sqrt{(LC)}}$

$$= \frac{1}{2\pi\sqrt{(5 \times 0.5 \times 10^{-6})}}$$

$$= 100.66 \text{ Hz}$$

(b) Q-factor

$$Q = \frac{1}{R}\sqrt{\frac{L}{C}}$$

$$= \frac{1}{150}\sqrt{\left(\frac{5}{0.5 \times 10^{-6}}\right)}$$

$$= 21.08$$

(c) When the frequency is twice the resonance frequency X_L is twice the value it has at resonance, X_C is half the value it has at resonance.

At resonance

$$X_L = 2\pi f_r L$$
$$= 2\pi \times 100.66 \times 5$$
$$= 3162.3 \ \Omega$$

At twice the resonant frequency

$$X_L = 2 \times 3162.3$$
$$= 6324.6 \ \Omega$$

At resonance

$$X_C = 1/2\pi f_r C$$
$$= 10^6/2\pi \times 100.66 \times 0.5$$
$$= 3162.3 \ \Omega$$

At twice the resonant frequency

$$X_C = 1/2 \times 3162.3$$
$$= 1581.1$$

So that the combined reactance

$$X_L - X_C = 4743.4 \ \Omega$$

and since

$$\tan \phi = \frac{X_L - X_C}{R}$$

where ϕ is the phase angle

$$\tan \phi = \frac{4743.4}{150}$$
$$= 31.62$$

and $\qquad \phi = 1.54$ rad from tables or calculator.

Specific objectives

The expected learning outcome is that the student:
2.11 *Draws the phasor diagram for a two-branch parallel circuit with C in one branch and only (a) L (b) L–R (c) R, in the other.*
2.12 *Determines the current in each branch of the circuit in 2.11.*
2.13 *Determines the sum of the branch currents using*
   (a) scaled phasor diagrams, and
   (b) resolution into in-phase and quadrature components.
2.14 *Determines the impedance of the circuit referred to in 2.11.*
2.15 *Determines the voltage, current, phase angle in the circuit in 2.11.*

Parallel circuits
C-R alone

Consider the circuit, shown in fig. 2.3 consisting of a capacitor of capacitance C in parallel with a resistor of resistance R. The supply voltage is shown as V_S, the supply current as I_S and the currents in the capacitive and resistive branches of the circuit as I_C and I_R respectively.

Since the circuit is a parallel circuit the whole of the supply voltage V_S is across the capacitor and across the resistor so that

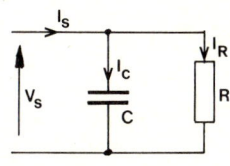

Figure 2.3

I_R is in phase with V_S
I_C is in quadrature (leading) with V_S

as shown in the phasor diagram in fig. 2.4.

I_C and I_R may be summed to give I_S, the phase angle ϕ being as shown.

Note that ϕ lies between $\pi/2$ (which it would be if the resistor were not present, i.e. R is infinite) and zero (which it would be if the capacitor is not present (i.e. X_C is infinite).

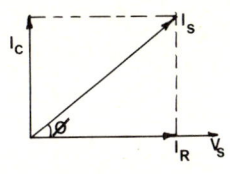

Figure 2.4

The circuit impedance

$$Z = V_S/I_S$$
$$\text{so that } I_S = V_S/Z$$

From the phasor diagram

$$I_S^2 = I_R^2 + I_C^2$$

and since

$$I_R = \frac{V_S}{R} \text{ and } I_C = \frac{V_S}{X_C}$$

then

$$\frac{V_S^2}{Z^2} = \frac{V_S^2}{R^2} + \frac{V_S^2}{X_C^2}$$

i.e.

$$\frac{1}{Z^2} = \frac{1}{R^2} + \frac{1}{X_C^2}$$

and

$$Z = \sqrt{\left[1/\left(\frac{1}{R^2} + \frac{1}{X_C^2}\right)\right]}$$

The phase angle ϕ may be obtained from

$$\cos \phi = I_R/I_S$$

$$= \frac{V_S \, Z}{R \, V_S}$$

$$= Z/R$$

or by

$$\sin \phi = I_C/I_S$$

$$= \frac{V_S \, Z}{X_C \, V_S}$$

$$= Z/X_C$$

or from

$$\tan \phi = I_C/I_R$$

$$= \frac{V_S \, R}{X_C \, V_S}$$

$$= R/X_C$$

(the reciprocal of the relationship in the series circuit containing R and C alone; in that circuit $\tan \phi = X_C/R$).

L-C alone   An inductor without resistance and having any substantial value of inductance is difficult if not impossible to arrange. Inductance, as we have seen, is a property of induction of e.m.f.s due to a magnetic field being set up by component current. To obtain a substantial value of inductance, wire is usually formed in the shape of a coil, the greater the number of turns the greater the inductance. Even if the wire is very thick and an extremely good conductor the coil will still have resistance. In order to understand the basic theory of a parallel circuit containing an inductor in one branch and a capacitor in the other, we shall start first, however, by ignoring the inductor resistance and consider its resistance alone. This assumption considerably simplifies the theory.

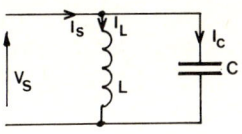

Figure 2.5

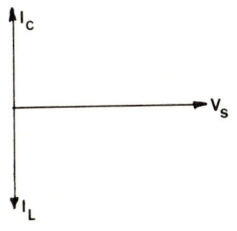

Figure 2.6

CIVIL

Fig. 2.5 shows the circuit diagram of such a circuit its phasor diagram being shown in fig. 2.6. The circuit consists of an inductor of inductance L and negligible resistance connected in parallel with a capacitor of capacitance C. Currents I_L, I_C flow in the inductive and capacitive branches respectively, the total current being I_S and the applied voltage shown as V_S.

Current I_C leads V_S by $\pi/2$ rad and current I_L lags V_S by $\pi/2$ rad, as shown in fig. 2.6. Current I_S is the phasor sum of I_L and I_S and since they are opposite in phase

$$I_S = I_L \sim I_C$$

(where $\sim$ again means 'the difference between').

There are three possible cases: I_L being larger than I_C, I_C being larger than I_L and the two being equal. In the first case

$$I_S = I_L - I_C$$

in the second

$$I_S = I_C - I_L$$

in the third

$$I_S = 0$$

The phasor diagrams are shown in fig. 2.7. Which phasor diagram applies to the circuit at any time depends upon the supply frequency, since this in turn determines the reactance of each branch of the circuit and thus the current flowing in it.

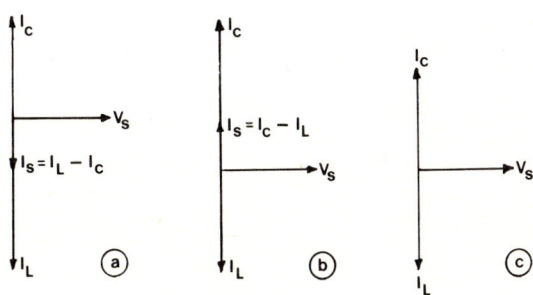

Figure 2.7

Specific objectives   *The expected learning outcome is that the student:*
2.16 *States the conditions for resonance in a parallel circuit with L and R in one branch and C only in the other.*
2.17 *Applies the formula for the parallel resonance frequency.*
2.18 *Defines Q-factor as the current magnification for the circuit 2.16.*

Parallel resonance   In part (c) of fig. 2.7 the two currents are equal and it follows that since the same voltage, V_S, is applied to each branch

$$X_C = X_L$$

The frequency at which this occurs is called the *resonant frequency* (as with the series circuit) and is denoted by f_r and the circuit is said to be in resonance. Since the circuit is parallel connected, *parallel resonance* is said to occur when the frequency is equal to the resonant frequency. Since

$$X_C = X_L$$

$$\frac{1}{2\pi f_r C} = 2\pi f_r L$$

and

$$f_r = \frac{1}{2\pi\sqrt{(LC)}}$$

as with the series circuit (but note that this formula is true here only because resistance has been neglected).

At resonance the supply current is zero despite there being a value of supply voltage and the total circuit impedance appears infinitely high.

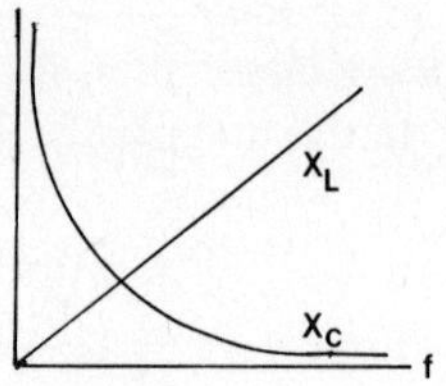

The remaining parts of fig. 2.7 show conditions below resonance (fig. 2.7a) and above resonance (fig. 2.7b). Below resonance, since $X_L = 2\pi fL$ and $X_C = 1/2\pi fC$ (see fig. 2.8):

X_C is larger than X_L
I_C is smaller than I_L
$I_S = I_L - I_C$

i.e. circuit overall is *inductive* since the supply current *lags* the supply voltage (Fig. 2.6a).

Above resonance

X_L is larger than X_C
I_L is smaller than I_C
$I_S = I_C - I_L$

and the circuit overall is *capacitive* since the supply currents *leads* the supply voltage (Fig. 2.7b).

This is the reverse of the series circuit. There at frequencies below resonance the circuit is capacitive and at frequencies above resonance the circuit is inductive.

Figure 2.8

LCR circuit   The practical *L—C—R* parallel circuit is shown in fig. 2.9 together with its phasor diagram.

The phasor diagram in fig. 2.9 should be studied carefully. The

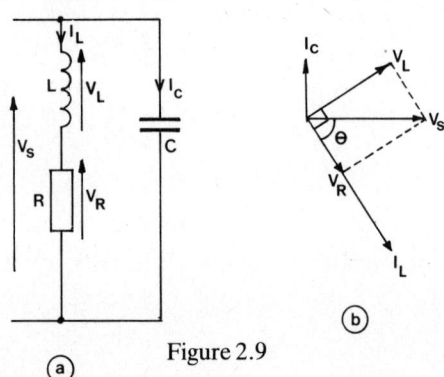

Figure 2.9

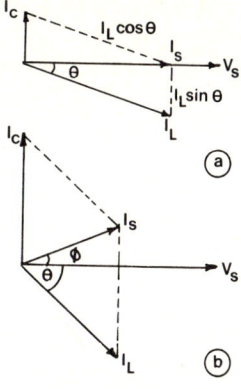

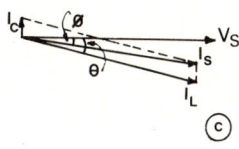

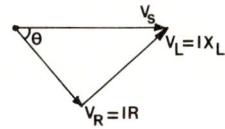

Figure 2.10

Figure 2.11

relative positions of the capacitor current I_C and the supply voltage V_S are as before (I_C leading V_S by $\pi/2$ rad). The inductor branch current I_L, however, no longer lags V_S by $\pi/2$ rad because of the effect of the inductor branch resistance R. The inductor branch is itself a series circuit containing inductance and resistance and the voltages across these components (or, to be exact, these parts of the *same* component, the inductor) must sum to the supply voltage, V_S. The relative positions of V_L and V_R and the size of *branch phase angle* Θ depend upon their magnitude which in turn depends upon frequency. The inductive branch current, I_L, is always, however, in phase with V_R wherever that may lie. As before there are three conditions of interest, resonance and frequencies below and above the resonant frequency. The phasor diagrams for these conditions are shown in fig. 2.10.

Fig. 2.10a shows conditions at resonance. The capacitive branch current I_C is in quadrature with V_S, leading it by $\pi/2$ rad. The inductive branch current I_L lags V_S by some angle Θ which is determined by the relative values of the inductive reactance X_L and resistance R. Fig. 2.11 shows the phasor diagram of this branch and from the diagram we see that

$$\tan \Theta = X_L/R$$

In fig. 2.10a the phasor sum of I_C and I_L, that is the supply current I_S, lies in phase with the supply voltage V_S and the circuit appears resistive. From fig. 2.10a it can be seen that $I_S = I_L \cos \Theta$, the other component of I_L which is $I_S \sin \Theta$ (shown dotted)being equal and opposite in phase to I_C. At frequencies above resonance the capacitive reactance (which equals $1/(2\pi fC)$) is lower in value, and I_C higher and the impedance of the inductive branch (which equals $\sqrt{(X_L{}^2 + R^2)}$ where $X_L = 2\pi fL$) is higher, making I_L smaller. The inductive branch phase angle Θ is increased and the phasor sum of I_C and I_L, that is, I_S, the supply current, now leads the supply voltage by some angle ϕ, the phase angle of the parallel circuit as a whole. The circuit above resonance is *capacitive* (a series circuit above resonance is inductive). See fig. 2.10b.

Conditions below resonance are shown in fig. 2.10c. Here phase angle Θ is lower than at resonance, I_C is smaller (since $X_C \propto 1/f$) and I_L is larger ($X_L \propto f$) and the supply current I_S now lags V_S. The circuit appears inductive (a series circuit is capacitive below resonance).

The parallel circuit containing inductance, capacitance and resistance behaves much as the (theoretical) parallel circuit containing inductance and capacitance only. Both circuits are inductive below resonance and capacitive above resonance and in both circuits the supply currents falls to a minimum at resonance (zero in the circuit containing no resistance), the circuit impedance being a maximum at the resonant frequency. As might be expected the relevant equations are more complicated for the practical circuit since we must now take resistance into account.

The resonant frequency f_r may be determined as follows.
At resonance

$$I_C = I_L \sin \Theta$$

and since

$$\sin \Theta = \frac{X_L}{\sqrt{(R^2 + X_L{}^2)}} \text{ (series } L\text{-}R \text{ circuit)}$$

$$I_C = \frac{I_L X_L}{\sqrt{(R^2 + X_L 2)}}$$

but

$$I_C = \frac{V_S}{X_C} \text{ and } I_L = \frac{V_S}{\sqrt{(R^2 + X_L{}^2)}}$$

so that

$$\frac{V_S}{X_C} = \frac{V_S}{\sqrt{(R^2 + X_L{}^2)}} \, \frac{X_L}{\sqrt{(R^2 + X_L{}^2)}}$$

and

$$\frac{1}{X_C} = \frac{X_L}{R^2 + X_L{}^2}$$

so that

$$R^2 + X_L{}^2 = X_L X_C \text{ and } X_L{}^2 = X_L X_C - R^2$$

and since, at resonance,

$$X_L = 2\pi f_r L \text{ and } X_C = \frac{1}{2\pi f_r C}$$

$$(2\pi f_r L)^2 = \frac{2\pi f_r L}{2\pi f_r C} - R^2$$

i.e. $f_r{}^2 (2\pi L)^2 = \dfrac{L}{C} - R^2$

Dividing through by $(2\pi L)^2$, i.e. $4\pi^2 L^2$

$$f_r{}^2 = \frac{L}{4\pi^2 L^2 C} - \frac{R^2}{4\pi^2 L^2}$$

$$= \frac{1}{4\pi^2 LC} - \frac{R^2}{4\pi^2 L^2}$$

$$= \frac{1}{4\pi^2} \left(\frac{1}{LC} - \frac{R^2}{L^2} \right)$$

so that

$$f_r = \frac{1}{2\pi} \sqrt{\left(\frac{1}{LC} - \frac{R^2}{L^2} \right)}$$

which is the equation for the resonant frequency.

$$\text{If } R = 0 \quad f_r = \frac{1}{2\pi \sqrt{(LC)}}$$

as was shown earlier. If R is relatively small this is the expression for the approximate value of the resonant frequency.

For the series L-C-R circuit the resonant frequency also has the value $1/[2\pi\sqrt{(LC)}]$ (whether resistance is present or not) and this expression is convenient to remember for both circuits, bearing in mind that for the practical parallel *L-C-R* circuit it is approximate only. In the practical parallel *L-C-R* circuit the value of the resistance *does* affect the value of the resonant frequency, however slightly this may be in practice.

The impedance of the parallel circuit may be derived as follows. At resonance

$$I_S = I_L \cos \Theta$$

$$= \frac{I_L R}{Z_L}$$

since $\cos \Theta = R/Z_L$ where Z_L is the impedance of the inductive branch (series *L-R* circuit). Now

$$I_L = \frac{V_S}{Z_L} \text{ and } I_S = \frac{V_S}{Z_r}$$

where Z_r is the impedance of the parallel circuit at resonance. So that

$$\frac{V_S}{Z_r} = \frac{V_S R}{Z_L^2}$$

$$\text{i.e. } Z_r = \frac{Z_L^2}{R}$$

$$= \frac{X_L^2 + R^2}{R}$$

(since $Z_L^2 = X_L^2 + R_2$)

Earlier, it was shown that,

$$R^2 + X_L^2 = X_L X_C$$

so that

$$X_L^2 = X_L X_C - R^2$$

and substituting in the above equation

$$Z_r = \frac{X_L X_C - R^2 + R^2}{R} = \frac{X_L X_C}{R}$$

$$= \frac{L}{CR} \quad \text{since } X_L = 2\pi f_r L$$
$$\text{and } X_C = 1/2\pi f_r C$$

$$Z_r = L/CR$$

This value of the circuit impedance at resonance is called the *dynamic impedance* of the parallel circuit.

Typical graphs of both impedance and current plotted against frequency for a practical parallel circuit are shown in fig. 2.12a. As is shown there the impedance rises as frequency is increased from zero until at resonance the peak value of the dynamic impedance is reached. As frequency is further increased the impedance falls in value. The current, being inversely proportional to impedance (for a constant suppy voltage), falls in value as frequency is increased

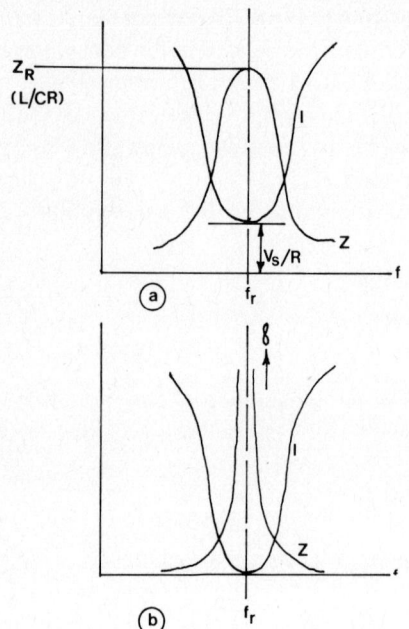

Z_R
(L/CR)

(a) f_r

(b) f_r

Figure 2.12

from zero and at resonance it has a minimum value. As frequency is further increased the current value rises again. If the resistance of the inductive branch is negligible the curves are modified as shown in fig. 2.12b. Note that if R is zero the dynamic impedance is infinitely high so that the supply current at resonance is zero.

The supply current to the circuit is always the phasor sum of the inductive branch current, I_L, and the capacitive branch current, I_C. Each of these will in itself be greater than the supply current since I_L acts either totally in opposition or partially in opposition (i.e. has a component in direction opposition) to I_C, depending upon whether R_L is negligible or otherwise (see phasor diagrams fig. 2.7c and 2.10a), and the phasor sum will be smaller than either of them. If R_L is negligible $I_C = I_L$ at resonance, the two are in antiphase and a circulating current is established in the circuit. The supply current is zero in this case and we have a situation similar to the 'voltage magnification' of the series circuit at resonance (in which reactive component voltages exceed the supply voltage). In this case the phenomenon is called *current magnification*. If R_L is not negligible current magnification still occurs since the current in either branch at resonance is larger than the current supplied although here the branch currents are not quite equal and opposite. The ratio between reactive component voltage and supply voltage at resonance in a series circuit, the voltage magnification or Q-factor, was earlier shown to have the value

$$Q = \frac{\omega L}{R} \text{ or } Q = \frac{1}{\omega CR}$$

In the parallel circuit,

$$I_L = \frac{V_S}{Z_L}$$

at resonance $V_S = I_S Z_r$

where I_S is the supply current and Z_r the dynamic impedance, so that

$$I_L = \frac{Z_r}{Z_L} I_S$$

The ratio Z_r/Z_L is the number of times that the current is greater than the supply current, i.e. the current magnification. Now

$$Z_r = \frac{L}{CR} \text{ and if } R_L \text{ is ignored } Z_L = X_L$$

so that

$$\frac{Z_r}{Z_L} = \frac{L}{CR} \frac{1}{X_L}$$

$$= \frac{L}{CR\omega L}$$

$$= 1/\omega CR$$

and since

$$I_L = \frac{Z_r}{Z_L} I_S \text{ then } I_L = QI_S$$

Example 2.3   A 50 mH, 40 Ω inductive coil is connected in parallel with a capacitor and the resonant frequency is found to be 10 kHz. Calculate:

(a) the capacitance of the capacitor;

(b) the dynamic impedance of the circuit;

(c) the capacitor current at resonance if the supply voltage is 25 V;

(d) the Q-factor of the circuit.

(a) The resonant frequency of the circuit is given by

$$f_r = \frac{1}{2\pi} \sqrt{\left(\frac{1}{LC} - \frac{R_L{}^2}{L^2}\right)}$$

Squaring both sides and rearranging

$$f_r{}^2 = \frac{1}{4\pi^2}\left(\frac{1}{LC} - \frac{R^2}{L^2}\right)$$

$$\frac{1}{4\pi^2 LC} = f_r{}^2 + \frac{R_L{}^2}{4\pi^2 L^2}$$

Multiplying throughout by $4\pi^2 L$

$$\frac{1}{C} = 4\pi^2 L f_r{}^2 + R_L{}^2/L$$

Inserting the given values

$$\frac{1}{C} = 4\pi^2 \times 50 \times 10^{-3} \times (10 \times 10^3)^2 + \frac{40^2}{50 \times 10^{-3}}$$

$$= 197.39 \times 10^6 + 3.2 \times 10^4$$
$$= 197.39 \times 10^6 + 0.032 \times 10^6$$
$$= 197.42 \times 10^6$$

and $C = 5.065 \times 10^{-9}$ F

The capacitance is 5.065 nF.

Notice that the second term (0.032×10^6) in the calculation, which is derived from $R_L{}^2/L^2$, may reasonably be ignored by comparison with the first term (197.39×10^6), which is derived from $1/LC$. The approximate equation

$$f_r = \frac{1}{2\pi\sqrt{(LC)}}$$

can be used with good accuracy here as is usually the case.

(b) The dynamic impedance of a parallel resonant circuit is given by

$$Z_r = \frac{L}{CR_L}$$

so that, here,

$$Z_r = \frac{50 \times 10^{-3}}{5.065 \times 10^{-9} \times 40}$$

$$= 246.8 \text{ k}\Omega$$

The dynamic impedance is 246.8 kΩ.

(c) The reactance of the capacitor at resonance is $1/2\pi f_r C$

$$\text{i.e. } X_C = \frac{1}{2\pi \times 10 \times 10^3 \times 5.065 \times 10^{-9}}$$

$$= 3142 \ \Omega$$

and the capacitor current is V_S/X_C where V_S is the supply voltage

$$\text{capacitor current} = \frac{25}{3142}$$

$$= 7.95 \times 10^{-3}$$

The capacitor current is 7.95 mA.

(d) $$Q = \frac{2\pi f_r L}{R}$$

$$= \frac{2\pi \times 10 \times 10^3 \times 50 \times 10^{-3}}{40}$$

$$= 78.54$$

The Q-factor is 78.54.

Example 2.4   The dynamic impedance of a parallel resonant circuit is 500 kΩ. The circuit consists of a 250 pF capacitor in parallel with a coil of resistance 10 Ω. Calculate the coil inductance, the resonant frequency and the Q-factor of the circuit.

Dynamic impedance

$$Z_r = \frac{L}{CR_L}$$

so that $$L = Z_r C R_L$$

$$= 500 \times 10^3 \times 250 \times 10^{-12} \times 10$$

$$= 1.25 \text{ mH}$$

Coil inductance is 1.25 mH.

The resonant frequency

$$f_r = \frac{1}{2\pi} \sqrt{\left(\frac{1}{LC} - \frac{R_L^2}{L^2}\right)}$$

$$= \frac{1}{2\pi} \sqrt{\left[\frac{1}{1.25 \times 10^{-3} \times 250 \times 10^{-12}} - \frac{10^2}{(1.25 \times 10^{-3})^2}\right]}$$

$$= 284.7 \text{ kHz}$$

The Q-factor

$$Q = 2\pi f_r L/R_L$$
$$= 2\pi \times 284.7 \times 10^3 \times 1.25 \times 10^{-3}/10$$
$$= 223.6$$

Specific objectives

The expected learning outcome is that the student:

2.21   *Defines the bandwidth.*

2.22   *Explains the effect of variation of component values upon bandwidth.*

2.23   *Draws response curves of simple coupled tuned circuits.*

2.24   *Describes the effect of variation of coupling upon response.*

2.25   *Explains the use of resonant circuits to select and amplify signals.*

Bandwidth

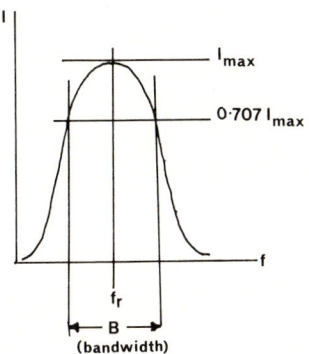

Figure 2.13

As we have seen, both series and parallel *L-C-R* circuits are frequency selective, that is, they respond differently as the frequency of the supply voltage is changed, a particular and unique response being obtained at one frequency, called the resonant frequency. The nature of the response depends upon whether the circuit components are connected in series or parallel.

In the series circuit the general shape of the graph of the circuit current plotted against frequency is as shown in fig. 2.13 (a graph of voltage developed across the resistor plotted against frequency has a similar shape).

As shown in the figure the curve is symmetrical about the centre or resonant frequency, the current rising to a peak value at this frequency. A useful parameter of the circuit when it is to be used in a frequency selective application, for example, in a radio receiver (considered in more detail shortly), is the *bandwidth*, symbol *B*, which is the separation between the frequencies at which the power developed by the circuit current falls to one half of the maximum value it has at resonance. The points at which this occurs are called the *half-power points*. The power developed at resonance is proportional to I_{max}^2 so that at the half-power points the power is proportional to $0.5\ I_{max}^2$, i.e. the current has fallen to a value

$$\sqrt{0.5\ I_{max}^2}$$

or $0.707\ I_{max}$, as shown in fig. 2.13.

A common application of resonant circuits is in the tuning of a radio receiver such that a single desired frequency is received with maximum response. Here the value of the bandwidth tells us the sharpness of the response of the circuit, i.e. is a measure of how particularly frequency selective the circuit is. A circuit having a large bandwidth offers a similar response to a number of frequencies around the resonant frequency, the response at resonance being only marginally different. If the bandwidth is reduced the response at resonance is markedly different to that at neighbouring frequencies and the circuit is very selective, a desirable feature when trying to tune a receiver to a particular signal frequency sited amongst a large number of other signal frequencies.

It can be shown that the bandwidth is dependent upon the *Q*-factor of the circuit the actual relationship being

$$B = f_r/Q$$

where f_r is the resonant frequency, and we see that as Q is increased the bandwidth is reduced and vice versa. A high Q circuit is thus extremely selective. Since the value of Q is given by

$$Q = \frac{\omega L}{R}$$

or

$$\frac{1}{\omega CR}$$

where L, C, R are the component values it can be seen that low values of inductance and high values of capacitance render low values of Q and reduce the selectivity. High values of inductance and low values of capacitance produce a high Q and high selectivity. See fig. 2.14a. The value of the resistance is assumed constant in the graphs shown in the figure; it too plays a part of course since, from the above relationships, Q is inversely proportional to R, a low value of resistance producing a high Q-factor and vice versa. See fig. 2.14b.

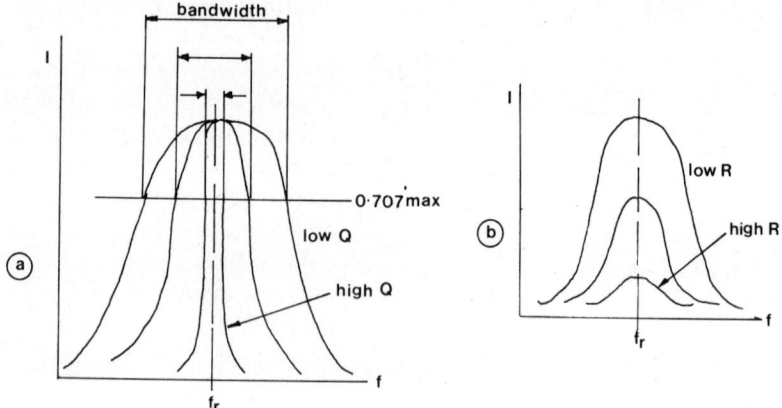

Figure 2.14

In the parallel resonant circuit similar curve shapes are obtained if the value of the circuit impedance is plotted against frequency, the maximum impedance, called the dynamic impedance Z_r being obtained at resonance. Here the bandwidth is the separation between the points at which the impedance falls to 0.707 of the maximum value, that is, the value of the dynamic impedance. Again the curve shape may be altered by changing component values, particularly that of the resistance, a shallow (low Q) curve being obtained by increasing the resistance value. Since

$$Z_r = \frac{L}{CR}$$

it can be seen that, other things being equal, increasing R reduces Z_r as shown in fig. 2.15. It should be noted that here altering the resistance value also affects the resonant frequency but the effect is very slight.

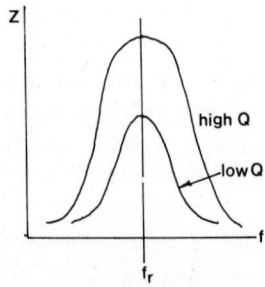

Figure 2.15

Coupled tuned circuits: use in r.f. amplifiers

Resonant circuits may be inductively coupled (via the coils) to give an overall response similar to a single parallel tuned circuit. The transformer action between the coils injects a voltage into the second circuit, fig. 2.16, so that it behaves in effect as a series circuit, and voltage magnification occurs across the capacitor. Used in this way to select radio frequency signals the signal is increased within the coupled circuits owing to:

(i) transformer action, and
(ii) voltage magnification

in addition to any amplification which occurs owing to the transistor of which the tuned circuit forms the load.

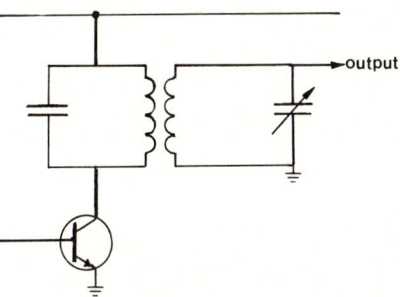

Figure 2.16

The bandwidth of a voltage amplifier is normally defined as the difference between the frequencies at which the voltage gain falls to 0.707 of the maximum gain. If two similar amplifiers are connected in cascade (the one feeding the other by mutually coupled tuned circuits) the overall gain will be approximately equal to the square of the individual gain. The overall gain at the frequencies where the single amplifier gain is 0.707 of the maximum will be 0.707×0.707, that is, 0.5 of the overall maximum. Thus the new overall bandwidth (which is the separation between the frequencies at which the overall gain falls to 0.707 of the overall maximum gain) will be *less* than the bandwidth of a single stage. With several stages the bandwidth may be quite severely reduced. This does of course considerably improve the frequency selecting abilities of the circuit (the selectivity) but may, if not closely watched, prove disadvantageous in that the communication signal (which consists of the radio frequency carrier frequency *plus* the frequencies of the conveyed intelligence) may be severely attenuated in some of its component frequencies. To avoid this the degree of coupling between amplifier stages is closely controlled to *increase* the bandwidth by 'flattening' the response curve slightly. One or more of the following methods may be used:

1. *Band-pass coupling* in which the degree of coupling is 'loosened' by altering the tuning of the coils by, say, movement of the 'slug' within the coil core (a threaded piece of ferromagnetic material which may be moved up and down the centre of the coil core).

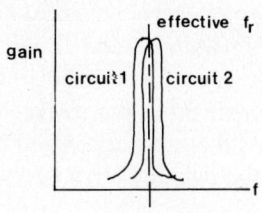

Figure 2.17

2. *Staggered tuning* in which the two tuned circuits are tuned to a slightly different frequency, the overall gain/frequency curve being flatter than either of the individual circuit curves and centred about a frequency midway between the individual circuit frequencies. See fig. 2.17.

3. *Damped tuning* in which each circuit is 'damped' by additional parallel resistors. The impedance at resonance is reduced by such a connection and the gain/frequency curve flattened slightly, thereby increasing the bandwidth.

Specific objectives

The expected learning outcome is that the student:
2.19   Determines the power dissipated in the circuits of 2.1 and 2.11.
2.20   Explains how the power-factor may be improved using static capacitors.

Power in parallel a.c. circuits

For the overall parallel circuit the same basic theory applies as to the series circuit namely, for a supply voltage V_S and supply current I_S phase displaced from V_S by angle ϕ

> power factor is $\cos \phi$
> power $P = V_S I_S \cos \phi$ (W)
> reactive volt-amperes $Q = V_S I_S \sin \phi$ (VAr)
> apparent power $S = V_S I_S$ (VA)

The power triangle shown in fig. 2.2c for the series circuit also applies here but in this case is not derived from the impedance triangle and the formulae given in the section on series circuits for S, P and Q in terms of the reactive and resistive components and the supply current do not apply since the supply current does not flow in each branch of the parallel circuit but is the phasor sum of the branch currents. Power calculations on parallel circuits are best carried out at this stage by reference to the phasor diagram. A method of simplifying calculations on a.c. circuits, especially parallel circuits, known as j-notation, is given in later units.

Example 2.5   A 10 H, 200 Ω inductor is connected in parallel with a 5 μF capacitor and the circuit then connected to a 240 V, 50 Hz supply. Calculate the supply current and the power factor of the circuit.

For the inductive branch of the circuit, the branch impedance

$$Z_L = \sqrt{(R_L{}^2 + X_L{}^2)}$$
$$= \sqrt{[200^2 + (2\pi \times 50 \times 10)^2]}$$
$$= 3148 \ \Omega$$

and the phase angle Θ is given by

$$\tan \Theta = X_L/R$$
$$= 2\pi \times 50 \times 10/200$$

from which $\Theta = 1.51$ rad.

The inductive branch current

$$I_L = \frac{V_S}{Z_L}$$

$$= \frac{240}{3148}$$

$$= 76.23 \text{ mA}$$

and lags the supply voltage by 1.51 rad.

The capacitive branch current

$$I_C = \frac{V_S}{X_C}$$

$$I_C = 240 \times 2\pi \times 50 \times 5 \times 10^{-6}$$

$$= 37.7 \text{ mA}$$

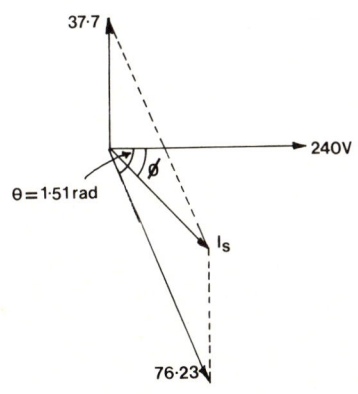

Figure 2.18

and leads the supply voltage by $\pi/2$ rad.

The phasor diagram may thus be drawn as shown in fig. 2.18 and we see that the supply current lags the supply voltage, the circuit being inductive overall at this frequency.

Calculation of I_S and cos ϕ may be carried out by applying trigonometric methods to the phasor diagram using sine and cosine rules or phasor resolution may be employed. The latter method is probably easier.

The inductive branch current 76.23 mA may be resolved into two components

76.23 cos Θ in phase with V_S
76.23 sin Θ lagging V_S by $\pi/2$ rad.

Horizontal component of this current is thus 76.23 cos (1.51 rad), i.e. 4.63 mA, and the vertical component is 76.23 sin (1.51 rad), i.e. 76.1 mA.

The quadrature (vertical) component of the supply current is the phasor sum of the capacitive branch current (37.7 mA) and the vertical component of the inductive branch current (76.1 mA), i.e. 76.1 − 37.7 which equals 38.4 mA and lags V_S by $\pi/2$ rad.

The in phase (horizontal) component of the supply current is the horizontal component of the inductive branch current, i.e. 4.63 mA, see fig. 2.19, so that the supply current I_S is given by

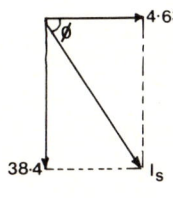

Figure 2.19

$$I_S = \sqrt{(38.4^2 + 4.63^2)}$$

$$= 38.7 \text{ mA}$$

and its phase angle ϕ is given by

$$\tan \phi = 38.4/4.63$$

so that $\phi = 1.45$ rad and the power factor cos $\phi = 0.12$ lagging.

The supply current is 38.7 mA lagging the supply voltage by 1.45 rad; the circuit power factor is 0.12.

Power factor improvement

In a practical situation where the load connected to the supply is perhaps one or more motors providing industrial drives for machinery, power factor becomes quite important. Electricity supply authorities charge for energy supplied, energy being the product of power and time, the unit commonly used being the kilowatt-hour (kWh).

Regardless of whether or not the power factor of a particular load is very low, the supply generators and supply cables must be constructed so that they are able to withstand both the supply voltage and the supply current even though the product of these two alone will not determine the revenue from which the cost of the supply network is met. For example, an alternator supply 11 kV at a rated current of 1000 A, say, must be capable of withstanding both the relatively high voltage and current and is capable therefore without further cost of providing

$$11 \times 10^3 \times 1 \times 10^3 \text{ i.e. } 11 \text{ MW}$$

of power. If the consumer is running a very inductive load with a power factor typically of 0.7, say, the power supplied (which determines the energy and the revenue) is only 0.7 × 11 MW, i.e. 7.7 MW. Supply authorities tend to penalise consumers who run low power factor loads by making an extra charge and it is therefore in the consumers' best interests to improve the load power factor as much as possible. Apart from this, of course, to some extent, the consumer has the same problem as the supplier in that the consumer network and equipment must be rated so as to be able to carry voltages and currents having products far higher than those of which practical use is being made. A fairly easy way to improve power factor of inductive loads, those most commonly met, is by the parallel connection of suitable capacitors.

Example 2.6 Calculate the true power, apparent power and reactive volt-amperes of the circuit given in Example 2.5 and determine the value of the capacitor required to raise the power factor to unity.

In the circuit of Example 2.5 the supply voltage is given as 240 V, the supply current and power factor were calculated to be 38.7 mA and 0.12 lagging, respectively.

$$\text{Apparent power } S = 240 \times 38.7 \times 10^{-3}$$
$$= 9.28 \text{ W}$$

$$\text{True power } P = S \cos \phi$$
$$= 9.28 \times 0.12$$
$$= 1.11 \text{ W}$$

$$\text{Reactive volt-amperes} = S \sin \phi$$
$$= 9.28 \times 0.993$$
$$= 9.21 \text{ W}$$

To change the power factor to unity an additional current must be taken from the supply which is equal and opposite to the quadrature component of the supply current, i.e. 38.4 mA, see the phasor diagram in fig. 2.19. This may be achieved by shunting the circuit as a whole by a capacitor, which will take this current from the supply.

Additional capacitor voltage is 240 V and its current must be 38.4 mA. Its reactance, therefore, is $240/(38.4 \times 10^{-3})$, i.e. 6250 Ω

and since the frequency is 50 Hz

$$6250 = \frac{1}{2\pi \times 50 \times C} \text{ where } C \text{ is the capacitance.}$$

Hence,

$$C = \frac{1}{6250 \times 2\pi \times 50}$$

$$= 0.51 \,\mu\text{F}$$

Addition of a 0.51 μF capacitor will change the overall circuit power factor to unity.

In this example the original circuit already contained some capacitance but overall behaved inductively because of the relative effect of the series inductor. In general, however, when circuits may not initially contain capacitance (as for example electric motors) the method remains the same. The quadrature (vertical) component of the supply current which is causing the current not to be in phase with the supply voltage is neutralised wholly or partly, depending upon whether the power factor is to be taken to unity or nearer to unity, by the addition of a parallel capacitor.

In practice capacitor banks may be used for power factor improvement or in some cases particular machinery which behaves capacitively under certain conditions may be added to the inductive load.

An example requiring partial improvement of power factor is provided at the end of the chapter, the solution also being provided.

Note that power factor improvement to unity is in effect creating the condition of resonance considered earlier in the chapter.

Example 2.7   A single phase induction motor takes 500 W from a 240 V, 50 Hz supply, the supply current being 3 A and lagging the supply voltage. Calculate the value of a single capacitor which, when connected in parallel with the machine, would improve the power factor to 0.9 lagging.

$$\text{Motor volt-amperes} = 3 \times 240 = 720 \text{ VA}$$
$$\text{Motor power} = 500 \text{ W}$$

$$\text{therefore power factor} = \frac{500}{720} = 0.69 \text{ lagging}$$

the phase angle being arc cos 0.69, i.e. 0.8 rad.

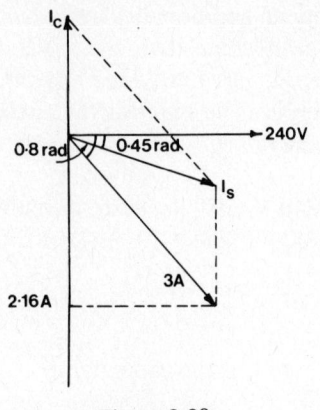

Figure 2.20

The phasor diagram is shown in fig. 2.20. The supply current of 3 A is shown lagging the supply voltage by 0.8 rad. When the capacitor is connected in parallel with the motor it will draw current such that a new supply current, the phasor sum of the capacitor current I_C and the 3 A taken by the motor, will flow. This is shown as I_S and lags the supply voltage by an angle having a cosine equal to 0.9, i.e. arc cos 0.9, which is 0.45 rad.

From the phasor diagram the horizontal component (in phase with the supply voltage) of the motor current (3 A) and the new supply current (I_S) are the same, i.e.

$$I_S \cos (0.45 \text{ rad}) = 3 \cos (0.8 \text{ rad})$$
$$\text{i.e. } 0.9 \, I_S = 0.69 \times 3$$
$$\text{and } I_S = 2.3 \text{ A}$$

The vertical component (in quadrature with the supply voltage) of the motor current is

$$3 \times \sin (0.8 \text{ rad}) = 2.16 \text{ A}$$

and of the new supply current is

$$2.3 \times \sin (0.45 \text{ rad}) = 1 \text{ A}$$

From fig. 2.20 the capacitor current I_C must equal the difference between these two quadrature components, i.e. 2.16 − 1 which is 1.16 A.
Now

$$\frac{240}{X_C} = 1.16 \text{ where } X_C \text{ is the reactance of the capacitor}$$

so that

$$X_C = \frac{1}{2\pi \times 50 \times C} = \frac{240}{1.16}$$

and

$$C = \frac{1.16}{-\pi \times 50 \times 240}$$

$$= 15.4 \, \mu\text{F}$$

In general terms the current drawn by the capacitor is always equal to the difference between the quadrature component of the load (motor) current at the 'old' power factor and the quadrature component of the 'new' supply current at the 'new' power factor. The 'new' supply current is the phasor sum of the 'old' supply current (the motor or load current) and the capacitor current.

Summary   For a series circuit $Z^2 = R^2 + (X_L \sim X_C)^2$ where Z is the impedance, R the resistance, X_L and X_C are the inductive and capacitive reactance, respectively.

$$X_L = 2\pi f L \text{ and } X_C = 1/2\pi f C$$

where f is frequency (Hz), L is inductance (H) and C is capacitance (F). If the phase angle is ϕ then

$$\sin \phi = X/Z; \cos \phi = R/Z \text{ and } \tan \phi = X/R$$

The circuit volt-amperes $S = V_S I_S$, reactive volt amperes $Q = V_X I_S$ or $V_S I_S \sin \phi$ and the circuit power $P = V_R I_S$ or $V_S I_S \cos \phi$ where V_S, I_S, V_X and V_R represent supply voltage, supply current, voltage across the reactance (i.e. $X_L \sim X_C$) and the voltage across the resistance. Cos ϕ is called power factor of the circuit.

When $X_L = X_C$ series resonance is said to occur, the frequency at which his occurs being denoted by f_r, given by

$$f_r = \frac{1}{2\pi\sqrt{(LC)}}$$

At resonance the voltage across X_L is equal to the voltage across X_C and each of these may be many times larger than V_R, the voltage across the resistance, and, since the reactive voltages are equal and opposite, also the supply voltage. Voltage magnification factor Q is given by

$$Q = \frac{V_L}{V_S} \text{ or } \frac{V_C}{V_S}$$

where V_L and V_C are the voltages across the inductance and capacitance, respectively, and also

$$Q = \frac{1}{R}\sqrt{\left(\frac{L}{C}\right)}$$

For a circuit containing a capacitor in parallel with a resistor only

circuit impedance $\qquad Z = \sqrt{\left[1/\left(\frac{1}{R^2} + \frac{1}{X_C^2}\right)\right]}$

the phase angle ϕ being obtained from

$$\cos \phi = \frac{Z}{R} \text{ or } \sin \phi = \frac{Z}{X_C} \text{ or } \tan \phi = \frac{R}{X_C}$$

For a circuit containing a pure inductance in parallel with a capacitor (not practicable but useful for analysis) the supply current is the phasor difference between the inductance current I_L and the capacitor current I_C, and, since these are in direct opposition, will lead the supply voltage by $\pi/2$ rad, lag the supply voltage by $\pi/2$ rad or be zero.

When $I_C = I_L$, parallel resonance is said to occur, the resonant frequency f_r being given by

$$f_r = \frac{1}{2\pi\sqrt{(LC)}}$$

as with the series circuit (but note this is because resistance has been ignored).

In a practical circuit containing inductance and resistance in one branch which is in parallel with a capacitor in the other branch, resonance occurs when the supply current, which is the phasor sum of I_C and I_L (the inductive–resistive branch current), lies in phase with the supply voltage. The resonant frequency

$$f_r = \frac{1}{2\pi} \sqrt{\left(\frac{1}{LC} - \frac{R^2}{L^2}\right)}$$

and the impedance of the circuit at resonance, Z_r, called the dynamic impedance, is given by

$$Z_r = \frac{L}{CR}$$

At resonance the current I_L may be many times larger than I_S, the supply current and the current magnification factor, Q, is given by the approximate relationship

$$Q = \frac{1}{\omega CR}$$

A series *L-C-R* circuit is capacitive below resonance, inductive above resonance; a parallel *L-C-R* circuit is inductive below resonance and capacitive above resonance.

The bandwidth B of a series *L-C-R* circuit is the separation in frequency between the frequencies at which the power developed by the circuit current falls to one half of the maximum value it attains at resonance. At these frequencies, called the half-power points, the current

$$I = 0.707\, I_{max}$$

Where I_{max} is the maximum value of circuit current. B, f_r and Q are related by the equation $B = f_r/Q$.

For a parallel *L-C-R* circuit the bandwidth is the separation between the frequencies at which the circuit impedance falls to 0.707 of the maximum or dynamic impedance Z_r attained at resonance.

The bandwidth and general shape of frequency response curves (i.e. current/frequency for a series circuit or impedance/frequency for a parallel circuit) may be considerably changed by altering component values R, L and C. When two tuned circuits are used in radio frequency amplifiers the overall frequency response may be too sharp and the bandwidth too small. This may be compensated for by band-pass coupling, staggered tuning or damped tuning.

Operating a.c. loads at low power factor is not economically sound since all cables and equipment must be rated to carry maximum values of voltage and current but the power obtained is not the voltage × current product but is equal to voltage × current × power factor. Power factor of inductive loads (which most industrial loads are) may be improved by connecting suitable capacitors in parallel with the load.

EXERCISE 2.1

1. A series *C-R* circuit of resistance 100 Ω has a phase angle of $\pi/3$ rad at 50 Hz. Calculate the value of the inductance placed in series with the circuit which would reduce the phase angle to zero.

2. When 60 V, 500 Hz is connected to a coil of resistance 50 Ω the supply current is 0.3 A. Calculate the value of series capacitance which would make the overall circuit resistive.

3. Two circuits connected in parallel across the same a.c. supply take 6 A, phase angle zero and 10 A, phase angle $\pi/6$ rad, respectively. Calculate the supply current and its phase angle.

4. The impedance of a parallel *L-C-R* circuit falls to 0.707 of the dynamic impedance at 750 Hz and 1250 Hz. Calculate the resonant frequency assuming a symmetrical curve and hence the *Q*-factor of the coil.

5. A 1 H, 200 Ω inductor is connected in parallel with a 8 μF capacitor across a 20 V variable frequency supply. Calculate
   (a) the frequency at which the supply current falls to a minimum
   (b) the value of the minimum supply current
   (c) the coil current when the supply current is minimum.

6. A reactive circuit is supplied with a 100 V alternating supply and a current of 2 A flows, phase angle $\pi/3$ rad. Calculate
   (a) the volt-amperes supplied
   (b) the power
   (c) the reactive volt-amperes.

7. The power factor of a 2 MVA load is to be improved from 0.8 lagging to 0.94 lagging. Calculate the kVAr required by the additional component.

8. A series *L-C-R* circuit of resistance 12 Ω and inductance 95.5 mH is connected to a 200 V, 60 Hz supply and a current of 10 A lagging flows. Calculate the reactance of the circuit capacitance.

Possible
marks

SELF-ASSESSMENT EXERCISE 2

1. State the formulae for inductive and capacitive reactance in terms of frequency component values. (3)

2. Sketch a typical graph of inductive and capacitive reactance plotted against frequency. (3)

3. State the relationships between *Q*-factor and the component values of a series *L-C-R* circuit (two required). (3)

4. Define 'dynamic impedance'. (3)

5. Give one reason why power factor improvement circuits are employed in industry. (3)

6. A series resonant circuit has a *Q*-factor of 100, an inductance of 10 H and a capacitance of 0.1 μF. Calculate the bandwidth of the circuit. (5)

7. A 1 μF, 100 kΩ series CR circuit is connected in parallel with a 0.2 H inductor of negligible resistance. Calculate the dynamic impedance of the parallel circuit. (5)

8. The voltage across the capacitor in a series *L-C-R* circuit at resonance is 200 V. The inductor has a reactance of 500 Ω and resistance of 50 Ω. Calculate the circuit supply voltage. (5)

9. A series circuit consisting of a 0.01 μF capacitor and a 2.54 μH, 2 Ω coil is connected across a 1 V a.c. supply. Calculate:
   (a) the resonant frequency;
   (b) the supply current at resonance;

(c) the voltage across the capacitor at resonance;
(d) the circuit Q-factor.   (14)

10. A 100/mH inductor of resistance 10 Ω has a Q-factor of 100. Calculate the bandwidth of a circuit containing this component in parallel with a capacitor. What is the approximate value of the capacitor?   (14)

11. A parallel resonant circuit contains a 10 mH, 50 Ω coil, the resonant frequency being 100 kHz. Calculate:
(a) the coil Q-factor;
(b) the dynamic impedance of the circuit;
(c) the capacitor current at resonance if the supply voltage is 100 V. (14)

12. A 1 H, 200 Ω inductor is connected to a 250 V, 50 Hz supply. Calculate:
(a) the supply current;
(b) the power factor;
(c) the power;
(d) the reactive volt-amperes;
(e) the value of the capacitor which when connected across the inductor increases the power factor to unity.   (14)

13. Describe what is meant by 'power factor improvement' in an industrial circuit and why it is economically desirable. Show with the aid of a phasor diagram how the value of the additional component may be calculated when it is desired to improve the power factor of an inductive load to a value other than unity.   (14)

Answers

SELF-ASSESSMENT EXERCISE 1

1. 0.55 H

2. 1.64 μF

3. 15.4 A, 0.33 rad

4. 1 kHz, 2

5. 56.4 Hz; 32 mA; 0.49 A

6. 200 VA; 100 W; 173.2 VAr

7. 619 kVAr

8. 20 Ω

SELF-ASSESSMENT EXERCISE 2

Marks

1. $X_L = 2\pi fL$; $X_C = 1/2\pi fC$ (1½ each)

2. As text; see fig. 2.8.   (1½ each)

3. $Q = \dfrac{\omega L}{R}$ and $Q = \dfrac{1}{\omega CR}$ (1½ each)

4. The dynamic impedance is the impedance of a parallel tuned circuit at resonance.   (1½)
In component value terms: $Z_r = L/CR$.   (1½)

5. Operation of equipment at low power factor is uneconomic.   (3)

6.
$$\text{Resonant frequency } f_r = \frac{1}{2\pi\sqrt{(LC)}}$$ (1½)

$$= \frac{1}{2\pi\sqrt{(10 \times 0.1 \times 10^{-6})}}$$

$$= 10^3/2\pi$$ (1)

$$\text{Since } Q = f_r/B, \ B = f_r/Q \qquad (1\tfrac{1}{2})$$

$$\text{bandwidth } B = \frac{10^3}{2\pi} \times \frac{1}{100}$$

$$= 10/2\pi$$

$$= 1.6 \qquad (1)$$

7.
$$Z_r = L/CR \qquad (2)$$
$$L = 0.2 \text{ H}$$
$$\text{and } CR = 1 \times 10^{-6} \times 100 \times 10^3 = 0.1 \qquad (2)$$
$$\text{Thus } Z_r = 0.2/0.1$$
$$= 2 \ \Omega \qquad (1)$$

8.
$$Q = \omega L/R \qquad (1\tfrac{1}{2})$$
$$= 500/50$$
$$= 10 \qquad (1)$$
$$V_C = 200$$
$$V_S = V_C/Q \qquad (1\tfrac{1}{2})$$
$$= 200/10$$
$$= 20 \qquad (1)$$

9. (a)
$$f_r = 1/2\pi\sqrt{(LC)} \qquad (2)$$

$$= \frac{1}{2\pi \times (2.54 \times 10^{-6} \times 0.01 \times 10^{-6})^{1/2}}$$

$$= 1 \text{ MHz} \qquad (1)$$

(b)
$$\text{Current at resonance} = V_S/R \qquad (2)$$
$$= 1/2$$
$$= 0.5 \text{ A} \qquad (1)$$

(c)
$$X_C = 1/2\pi fC \qquad (2)$$
$$= 1/2\pi \times 10^6 \times 0.01 \times 10^{-6}$$
$$= 15.9 \ \Omega \qquad (1)$$
$$V_C = X_C \times \text{current} \qquad (1)$$
$$= 0.5 \times 15.9$$
$$= 7.95 \text{ V} \qquad (1)$$

(d)
$$Q = V_C/V_S \text{ at resonance} \qquad (1)$$
$$= 7.95/1$$
$$= 7.95 \qquad (1)$$
$$(\text{alternatively } Q = \omega L/R \text{ or } 1/\omega CR)$$

10.
$$Q = \frac{\omega_r L}{R_L} \qquad (2)$$

$$\omega_r = \frac{QR_L}{L} \qquad (2)$$

$$\text{and } f_r = \frac{QR_L}{2\pi L} \qquad (2)$$

$$= \frac{100 \times 10}{2\pi \times 100 \times 10^{-3}}$$

$$= 1.59 \text{ kHz} \qquad (2)$$

$$B = \frac{f_r}{Q} = 15.9 \text{ Hz} \qquad (2)$$

$$f_r \simeq \frac{1}{2\pi\sqrt{(LC)}} \text{ i.e. } f_r^2 = \frac{1}{4\pi^2 LC} \qquad (2)$$

$$C = \frac{1}{4\pi^2 f_r^2 L} = \frac{1000}{4\pi^2 \times 1.59 \times 10^6 \times 100}$$

$$= 0.1 \ \mu F \tag{2}$$

11. (a) $$Q = \frac{2\pi f_r L}{R_L} \tag{2}$$

$$= \frac{2\pi \times 10^5 \times 10 \times 10^{-3}}{50}$$

$$= 125.7 \tag{1}$$

(b) $$Z_r = \frac{L}{C R_L} \tag{2}$$

The value of C is required. Neglecting R_L

$$f_r = \frac{1}{2\pi\sqrt{(LC)}} \tag{2}$$

and

$$C = \frac{1}{4\pi^2 f_r^2 L} \tag{1}$$

$$= \frac{1}{4\pi^2 \times 10^{10} \times 10 \times 10^{-3}}$$

$$= 253.3 \ pF \tag{1}$$

hence $$Z_r = \frac{10 \times 10^{-3}}{253.3 \times 10^{-12} \times 50}$$

$$= 789.6 \ k\Omega \tag{1}$$

(c) $$I_C = Q I_S \tag{2}$$
$$I_S = V_S / Z_r \tag{1}$$

$$= \frac{100}{789.6} \quad mA \tag{1}$$

$$I_C = \frac{125.7 \times 100}{789.6}$$

$$= 15.9 \ mA \tag{1}$$

12. (a) $$X_L = 2\pi \times 50 \times 1$$
$$= 314.2 \ \Omega \tag{1}$$

$$Z = \sqrt{(200^2 + 314.2^2)}$$
$$= 360 \ \Omega \tag{1}$$

Supply current $I_S = 250/360$
$$I_S = 0.694 \ A \tag{1}$$

(b) Phase angle $\phi = \arctan(X_L/R)$ $\qquad$ (1)
$$= \arctan(314.2/200)$$
$$\phi = 1 \ rad$$
$$\cos \phi = 0.537 \tag{1}$$

(c) Power $= V_S I_S \cos \phi$ $\qquad$ (1)
$$= 250 \times 0.694 \times 0.537$$
$$= 93.17 \ W \tag{1}$$

(d) Reactive volt amperes $= V_S I_S \sin \phi$ $\qquad$ (1)
$$= 250 \times 0.694 \times 0.841$$
(where $0.841 = \sin \phi$)
$$= 145.6 \ VAr \tag{1}$$

(e) Reactive current component $= I_S \sin \phi$ (1)
$$= 0.694 \times 0.841$$
$$= 0.584 \text{ A} \qquad\qquad (1)$$

This is the capacitor current if the reactive resultant is to be zero as the result of improving the power factor to unity.

Hence $X_C = 250/0.584$ (1)
$$= 428.3 \ \Omega$$

and $C = 1/2\pi f \, X_C$ (1)
$$= 1/(2\pi \times 50 \times 428.3)$$
$$= 7.43 \ \mu\text{F} \qquad\qquad (1)$$

13. For definition see text (must include addition of reactive component in parallel to compensate in whole or in part for reactive current of load. Reasons must include fact that cables and equipment must carry maximum voltage and current irrespective of power factor and penalty set by supply authority).   (5)

Phasor diagram: see fig. 2.20   (4)
and associated text and theory.   (5)

3 Three-phase supply

Topic area: C

General objective
The expected learning outcome is that the student identifies the basic principles of three-phase systems.

Specific objectives
The expected learning outcome is that the student:

4.1   *States the nature of, and reasons for, a three-phase supply network, with reference to the National Grid distribution system.*

4.2   *States the need for star and delta connections for power distribution.*

4.3   *Distinguishes between delta and star (3- and 4-wire) methods of connection.*

4.4   *States that under balanced load conditions:*
   (i) $V_L = V_{ph}$, $I_L = \sqrt{3}I_{ph}$ for a delta connection and,
   (ii) $I_L = I_{ph}$, $V_L = \sqrt{3}V_{ph}$ for a star connection.

4.5   *Performs calculations concerned with 4.4.*

4.6   *States that the power dissipation in a three-phase load is the sum of the single phase powers and that the power in a balanced three phase load is $\sqrt{3}V_L\,I_L\,\cos\phi$*

4.7   *Measures power in balanced and unbalanced three-phase loads using one, two and three wattmeters.*

4.8   *Draws phasor diagrams to identify the terms symmetrical and balanced.*

4.9   *Shows by phasor diagrams that the sum of line or phase currents in a balanced system is zero.*

4.10   *Draws phasor diagrams to explain normal and reverse phase rotation.*

4.11   *Calculates neutral current in a simple unbalanced 4-line system.*

4.12   *Shows by phasor diagrams that circulating voltage is twice line voltage in an incorrectly closed delta system.*

Electricity supply in the U.K.

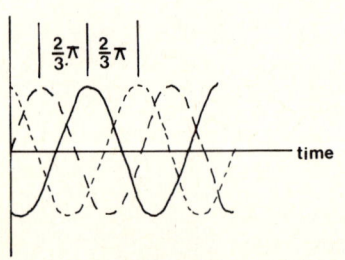

Figure 3.1

In the United Kingdom a *multiphase* system is used for the generation, transmission and distribution of alternating current. Three separate voltages of equal amplitude and frequency but separated in phase by 120 electrical degrees ($2\pi/3$ radians) are generated simultaneously as shown in fig. 3.1.

Each voltage is called a *phase* of the supply, the supply as a whole being called three phase, and each phase is distinguished from the other two by referring to it by number (1, 2 or 3) or by colour (red, yellow or blue).

A three-phase system of generation, transmission and distribution has the following advantages:

1. higher efficiency than a single-phase system;
2. availability of both industrial (three-phase) and domestic (single-phase) supplies from the same initial source;
3. the characteristics of three-phase machines are superior to those of single-phase machines: they include higher efficiency and power factor, less copper required per unit power, more even torque, self-starting.

Methods of connection

Since two conductors are normally required for the transmission of an electrical supply it may be thought that six conductors are required for a three-phase supply, since this is in effect carrying three single-phase supplies. While it is true that six conductors can be used it so happens that it is unnecessary because of the phase displacement chosen. Either three wires or four wires may be used depending upon the method of connection of the windings at the generator. There are two methods of connection of windings called *delta* or mesh connection and *star* connection. As with the generator windings, transformers and loads connected to the three-phase supply may also be connected in delta or star. See fig. 3.2.

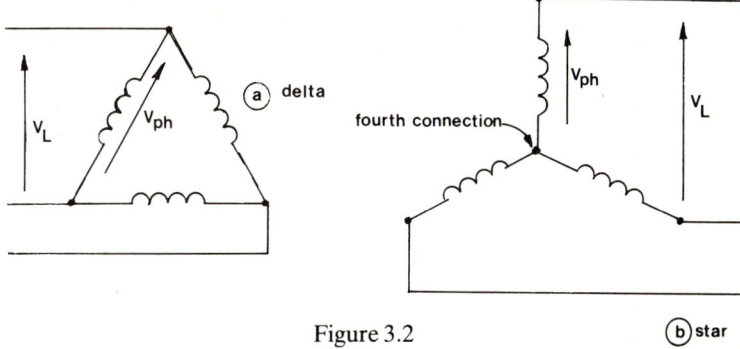

Figure 3.2

(a) delta

(b) star

Delta connection of an unloaded generator

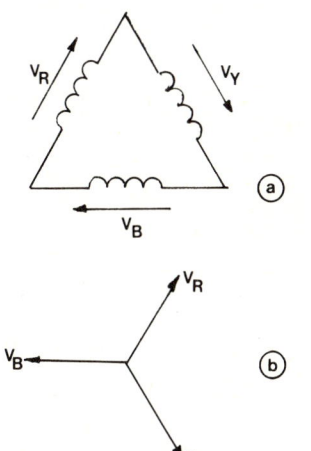

Figure 3.3

In the delta or mesh connection the three windings are connected in a closed loop as shown in fig. 3.3a. In each winding the voltage acts first in one direction and then the other, called for convenience, positive and negative, and the ends of each winding are connected so that when the voltage is positive it acts in the *same direction* in the loop, shown as a clockwise direction in the figure. It must be emphasised that not all three voltages are positive at any one time, for as fig. 3.1 shows, when one is positive, the other two might both be negative or one positive and one negative (the positive direction on a waveform graph usually being taken as above the time axis). However, when each voltage is positive in turn, it must act in the same loop direction for the following relationships to be true and to enable the mesh to be closed without ill effect.

If the equation yielding the instantaneous value of the red phase, V_R, is

$$V_R = V_{R\,max} \sin \omega t$$

Where $V_{R\,max}$ is the maximum voltage of the red phase then since the yellow phase voltage lags the red by $2\pi/3$ rad and the blue phase voltage lags the yellow by $2\pi/3$ rad, i.e. lags the red by $4\pi/3$ rad, we can write

$$V_Y = V_{Y\,max} \sin\,(\omega t - 2\pi/3)$$

$$V_B = V_{B\,max} \sin\,(\omega t - 4\pi/3)$$

where the subscript 'max' indicates maximum value of V_Y and V_B, respectively, V_Y and V_B being the instantaneous values of the yellow and blue phase voltages, respectively.

Provided that the windings are connected as described the resultant loop voltage is given by

$$V_{R\,max} \sin\,\omega t + V_{Y\,max} \sin\,(\omega t - 2\pi/3) + V_{B\,max} \sin\,(\omega t - 4\pi/3)$$

which on expansion gives

$$V_{R\,max} \sin\,\omega t + V_{Y\,max} \sin\,\omega t \cos 2\pi/3 + V_{Y\,max} \cos\,\omega t \sin 2\pi/3$$
$$+ V_{B\,max} \sin\,\omega t \cos 4\pi/3 + V_{B\,max} \cos\,\omega t \sin 4\pi/3$$

i.e. $V_{R\,max} \sin\,\omega t - \dfrac{1}{2}V_{Y\,max} \sin\,\omega t + \dfrac{\sqrt{3}}{2} V_{Y\,max} \cos\,\omega t - \dfrac{1}{2}V_{B\,max}$

$\sin\,\omega t - \dfrac{\sqrt{3}}{2} V_{B\,max} \cos\,\omega t$

and if $V_{R\,max} = V_{Y\,max} = V_{B\,max}$ as stated earlier, the first term is equal and opposite to the sum of the second and fourth terms and the third and fifth terms are equal and opposite so that the expression and therefore the resultant loop voltage is zero. Note that if any one of the windings is reversed, so that when its voltage is positive it does not act in the same loop direction as the other voltages when positive, the first expression (before expansion) will not contain two plus signs and the expansion does not yield zero.

The connections to the mesh (at the generator) from which the three-phase supply is taken are made at each junction of two windings to give three supply *lines* as shown in fig. 3.2a. As can be seen, the voltage between any two supply lines, called the *line voltage*, is equal to the phase voltage. Denoting line voltage by V_L and phase voltage by V_{ph} then

$$V_L = V_{ph}$$

The current in any line, however, is equal to the phasor sum of the phase currents in each of the two phase windings connected to the supply point from which the line is taken.

Effect of the load

The magnitude and phase relationships of the currents flowing in the generator windings are determined by the nature of the load which is connected to the supply. The load generally consists of three separated loads connected either in delta or star and overall may be *balanced* or *unbalanced*.

A balanced load is one in which each part of the total load has exactly the same impedance (both in overall magnitude and in the resistive and reactive components making up the impedance); an unbalanced load consists of three parts which are not identical and differing perhaps in overall magnitude, in the phase angle introduced between voltage and current or in both.

For a balanced load each phase of the supply is affected in exactly the same way so that the voltage relationships derived earlier still apply, i.e. the terminal voltages supplied per phase are equal in magnitude, each one is displaced from the next in phase by $2\pi/3$ rad, and the line voltage and phase voltage are equal.

Mesh connection currents with a balanced load

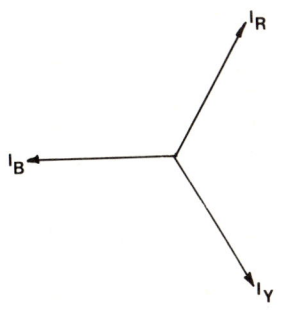

Figure 3.4

Assuming a balanced load the currents in each phase will be equal in magnitude and each current will be displaced from its neighbour by $2\pi/3$ rad as are the voltages. Each phase current may or may not be in phase with the corresponding phase voltage depending upon whether or not each limb of the load is purely resistive, but either way the phase angle between phase voltage and phase current is the same in each phase when the load is balanced so that the phasor diagram for the phase currents will appear as in fig. 3.4 and may be in line with or displaced from the voltage phasor diagram as shown in fig. 3.3b.

Denoting the line current in the line connected to the common point G of the red and blue phase windings by I_{RB}, as shown in fig. 3.5. we can see that I_{RB} is the *phasor difference* between the phase currents I_R and I_B since the 'positive' direction for I_R is towards

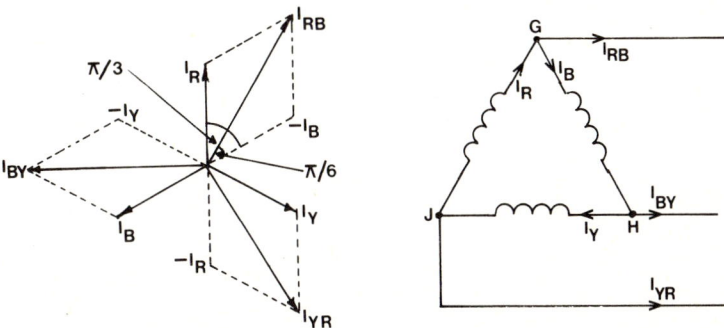

Figure 3.5

point G and for I_B is away from point G. On occasion, of course, I_R and I_B may be flowing towards the point and therefore adding together to give the line current at that particular instant but this still represents a *phasor* difference overall since if

$$I_{RB} = I_R - I_B \ (I_B \text{ flowing from point G})$$

then when I_B flows towards point G it must be written as $- I_B$ and

$$I_{RB} = I_R - (-I_B)$$
$$= I_R + I_B$$

assuming I_R to be flowing *towards* G and therefore shown as *positive* on both these occasions.

In fig. 3.5 to determine the value of I_{RB} if

$$\overline{I_{RB}} = \overline{I_R} - \overline{I_B}$$

$-I_B$ is drawn equal and opposite to I_B as shown and the phase parallelogram is constructed. Since I_R and I_B are equal in magnitude for a balanced load, I_{RB} will lie midway between I_R and $-I_B$ and since these are displaced from another by $\frac{1}{2} \times 2\pi/3$ rad, i.e. $\pi/3$ rad (see fig. 3.5) I_{RB} is displaced from I_R and I_B by $\frac{1}{2} \times \pi/3$ rad, i.e. $\pi/6$ rad so that by resolution

$$I_{RB} = I_R \cos \pi/6 + I_B \cos \pi/6$$
$$= \frac{\sqrt{3}}{2} I_R + \frac{\sqrt{3}}{2} I_B$$

Denoting I_{RB} by I_L (the line current) and $I_R (= I_B)$ by I_{ph} (the phase current)

$$I_L = \sqrt{3}\, I_{ph}$$

which is the general equation connecting line and phase currents for a delta connected supply assuming a balanced load.

Star connection of an unloaded generator

In star connection the windings are joined together at one point, called the *star* or *neutral* point, as shown in fig. 3.6 to the supply output lines. The connection is such that the voltage in each winding, that is, the phase voltage, acts outwards from the star point when positive.

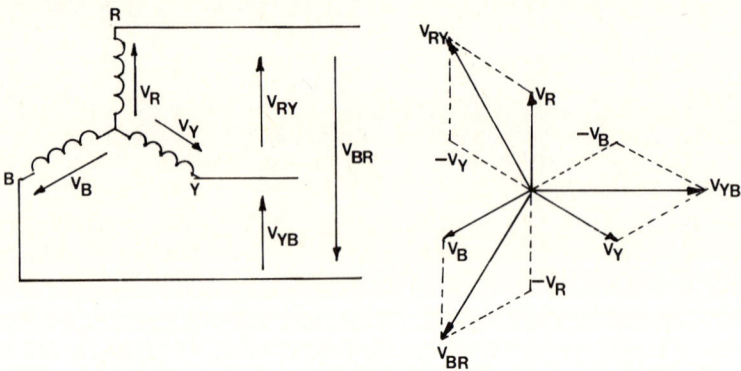

Figure 3.6

Here the line voltage is the phasor difference between the two phase voltages, the point concerning instantaneous directions being as explained when currents in the delta system were considered.

In the diagram

$$V_{RY} = V_R - V_Y$$
$$V_{YB} = V_Y - V_B$$
$$V_{BR} = V_B - V_R$$

The phase displacement between V_R and $-V_Y$ is $\pi/3$ rad so that

$$V_{RY} = V_R \cos \pi/6 + V_Y \cos \pi/6$$

and if $V_R = V_Y$

$$V_{RY} = \sqrt{3}\, V_R \text{ or } \sqrt{3}\, V_Y$$

Similarly

$$V_{YB} = \sqrt{3}\, V_Y \text{ or } \sqrt{3}\, V_B$$

and

$$V_{BR} = \sqrt{3}\, V_B \text{ or } \sqrt{3}\, V_R$$

or in general

$$V_L = \sqrt{3}\, V_{ph}$$

Effect of the load

When a load is connected to the generator, line and phase currents flow. It can be seen from fig. 3.6 that whether or not the load is balanced the line current is equal to the phase current *in the phase to which the line is connected*. For a balanced load the phase currents are all equal to each other in magnitude and phase as are the line currents. For an unbalanced load the phase currents may not be equal to each other but, as stated, the line current in the line connected to a particular phase, is equal to the phase current in that phase.

For a balanced load the phase voltages are all equal to each other and the line voltage

$$V_L = \sqrt{3}\, V_{ph}$$

as shown. For an unbalanced load this may not be the case and is considered later.

The neutral or star point connection

With the star connection there is a fourth point from which a supply line may be taken, the centre or star point. We can therefore have a three-phase, three-wire or a three-phase, four-wire system of supply. The first type of system is useful for balanced loads requiring a three-phase supply and is commonly used for industrial supplies. The second type of system using a fourth wire is used to provide three separate single phase supplies from a three-phase generator and is best employed when the load is likely to be unbalanced, as in the case of domestic supplies. Here the neutral point of the generator is connected to the neutral point of the final distribution transformer, the three separate supplies being taken from the transformer secondary. Commonly, on a housing estate, for example, one or more streets are supplied from a single phase from the distribution sub-station, all three phases being used separately. Under these conditions the three-phase load as a whole is almost invariably unbalanced.

Example 3.1 A three-phase star connected generator supplies a balanced delta connected load. If the voltage and current per phase of the generator are respectively 230 V and 25 A calculate the :
  (a) line voltage and current of the supply;
  (b) line voltage and current of the load;
  (c) phase voltage and current of the load.

(a) for the generator phase voltage = 230 V, phase current = 25 A.

Since it is star connected

$$\text{line voltage} = \sqrt{3} \times \text{phase voltage}$$
$$= \sqrt{3} \times 230$$
$$= 398.4 \text{ V}$$

$$\text{and line current} = \text{phase current}$$
$$= 25 \text{ A}$$

(b) For the load line voltage and current are clearly the same as for the generator, i.e. 398.4 V and 25 A.

(c) The load is delta connected so that

$$\text{phase voltage} = \text{line voltage}$$
$$= 398.4 \text{ V}$$

and the phase current

$$= \frac{1}{\sqrt{3}} \times \text{line current}$$
$$= 25/\sqrt{3}$$
$$= 14.4 \text{ A}$$

Specific objectives

The expected learning outcome is that the student:
4.6 *States that the power dissipation in a three-phase load is the sum of the single-phase powers and that the power in a balanced three-phase load is $\sqrt{3}V_L I_L \cos \phi$.*
4.7 *Measures power in balanced and unbalanced three-phase loads using one, two and three wattmeters.*

Power in a balanced three-phase load

For a balanced load the line voltages are equal to each other as are the phase voltages. Similarly the current in each line is the same and the phase currents are equal to each other. The phase displacement in each part of the load is the same so that the power factor in each phase is the same.

Denoting the line voltage by V_L, the phase voltage by V_{ph}, the line current by I_L, the phase current by I_{ph} (all r.m.s. values) and the phase angle and power factor by ϕ and $\cos \phi$, respectively.

$$\text{power per phase} = V_{ph} I_{ph} \cos \phi$$
$$\text{and the total power} = 3 V_{ph} I_{ph} \cos \phi$$

For a star system

$$V_L = \sqrt{3} V_{ph}$$
$$I_L = I_{ph}$$

and

$$\text{total power} = 3 \frac{V_L}{\sqrt{3}} I_L \cos \phi$$
$$= \sqrt{3} V_L I_L \cos \phi$$

For a delta system

$$V_L = V_{ph}$$
$$I_L = \sqrt{3} I_{ph}$$

$$\text{and total power} = 3V_L \frac{I_L}{\sqrt{3}} \cos \phi$$

$$= \sqrt{3} \, V_L \, I_L \cos \phi \text{ as before}$$

i.e. the expression for total power is the same regardless of the method of load connection provided that the system is balanced.

Example 3.2   Find the phase currents when three 10 μF capacitors are star connected across a 400 V, 50 Hz three-phase supply.

$$\text{The reactance per phase} = 10^6/2\pi \times 50 \times 10$$
$$= 318.2 \, \Omega$$
$$\text{Phase voltage} = 400/\sqrt{3}$$
$$= 230.9$$
$$\text{Phase current} = 230.9/318.2$$
$$= 0.725 \text{ A}$$

Example 3.3   Three identical resistors are star connected across a 440 V, three-phase supply and a line current of 4 A flows. Calculate:

    (a) the value of each resistor;
    (b) the power consumed.

(a)
$$\text{Phase voltage} = 440/\sqrt{3}$$
$$= 254 \text{ V}$$
$$\text{Phase current} = \text{line current}$$
$$= 4 \text{ A}$$
$$\text{Therefore load resistance} = 254/4 \text{ per phase}$$
$$= 63.5 \, \Omega$$

(b)
$$\text{Power} = \sqrt{3} \times 440 \times 4$$
$$= 3048.4 \text{ W}$$

Example 3.4   A three-phase delta connected motor connected to a 400 V, 50 Hz supply has an output of 29.84 kW, the efficiency and power factor being 95% and 0.9, respectively. Calculate the phase current in each winding.

$$\text{Motor input power} = \frac{29\,840}{0.95}$$
$$= 31\,410 \text{ W}$$

$$\text{Input power} = \sqrt{3} \, V_L \, I_L \cos \phi$$
$$= \sqrt{3} \times 400 \times I_L \times 0.9$$
$$\text{so that } 31\,410 = \sqrt{3} \times 400 \times I_L \times 0.9$$

$$\text{and } I_L = 50.37 \text{ A}$$

$$\text{Phase current} = 50.37/\sqrt{3}$$
$$= 29.08 \text{ A}$$

Example 3.5   A three-phase 500 V star connected generator supplies 74.6 kW to a delta connected motor at a power factor of 0.85 lagging. The motor efficiency is 0.9 p.u. Calculate:

(a) the generator phase current;
(b) the motor phase current.

(a) $$\text{Motor input} = 74\,600/0.9$$
$$= 82\,889\ \text{W}$$

so that $\sqrt{3} \times 500 \times I_L \times 0.85 = 82\,889$

and hence $I_L = 112.6\ \text{A}$

Since the generator is star connected

$$I_{ph} = I_L = 112.6\ \text{A}$$

(b) The motor is delta connected so that

$$\text{phase current} = I_L/\sqrt{3}$$
$$= 112.6/\sqrt{3}$$
$$= 65\ \text{A}$$

Measurement of power

Power in a three-phase system is measured using one or more wattmeters suitably connected. The number of wattmeters and the method of connection depends largely on, first, whether or not the load is balanced and, secondly, whether or not the neutral point, if there is one, is readily accessible.

Single wattmeter method

In all cases the total power is the sum of the power per phase of the load but determination of the power per phase may not be readily obtained using a single wattmeter per phase since the neutral point, if there is one, may not be accessible. For a star connected balanced load a single wattmeter may be used as shown in fig. 3.7, in which case the wattmeter reading, which is equal to

$$V_{ph}\,I_{ph}\cos\phi$$

is multiplied by three to give the total power absorbed by the load.

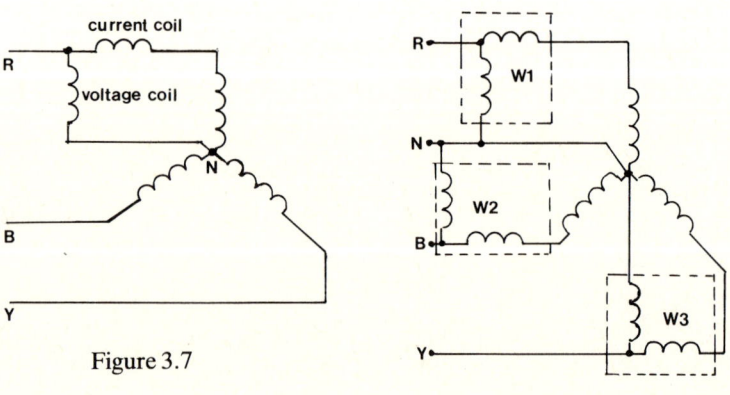

Figure 3.7

Figure 3.8

Three wattmeter method

For a star connected unbalanced load, three wattmeters each connected to a separate phase may be used and the total power then is the sum of the three readings of the wattmeters. See fig. 3.8.

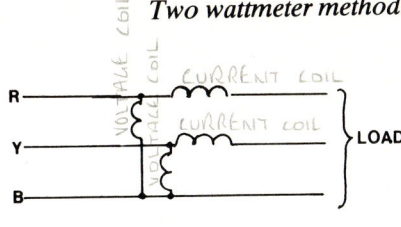

Two wattmeter method

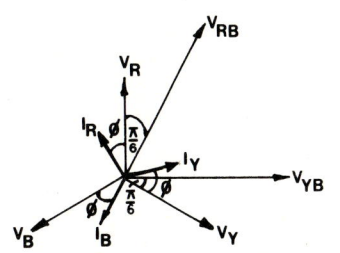

(a)

(b)

Figure 3.9

The most commonly used method for measuring power in a three-phase system regardless of the state of balance, waveform, phase sequence or load connection (star or delta) or whether or not the star point, if there is one, is readily accessible is the two wattmeter method as shown in fig. 3.9a.

In this method the wattmeter current coils are connected in any two lines, the voltage coils being connected between each of these two lines and the third line. As stated, the sum of the readings obtained is the total load power regardless of the type of load or its state of balance. As can be seen in the figure the star point is not required so its presence or accessibility is not relevant.

The formal analysis for the particular case of a balanced load and sinusoidal waveform follows, the phasor diagram being shown in fig. 3.9b.

In the figure V_R, V_Y and V_B represent the r.m.s. values of voltage per phase, the voltages between lines being

V_{RB} between red and blue lines
V_{YB} between yellow and blue lines

The line currents (i.e. those in the wattmeter current coils) are shown as I_R, I_B and I_Y, respectively. The phase angle per phase is shown s ϕ and is leading.

For wattmeter W1 the power reading is given by

$$P_1 = V_{RB} I_R \cos (\phi + \pi/6)$$

and for wattmeter W2 is given by

$$P_2 = V_{YB} I_Y \cos (\phi - \pi/6)$$

In both cases if ϕ is a lagging phase angle it becomes negative in these expressions and since $\cos \phi$ and $\cos (-\phi)$ are equal for values of ϕ from zero to π (the limits of phase difference between voltage and current of regular waveform) there is no effect on the expansion of the expressions which follow.

Since the system is balanced we may write the line voltage

$$V_L = V_{RB} = V_{YB} \text{ and line current } I_L = I_R = I_Y$$

so that

$$\begin{aligned}
P_1 &= V_L I_L \cos (\phi + \pi/6) \\
&= V_L I_L (\cos \phi \cos \pi/6 - \sin \phi \sin \pi/6) \\
&= V_L I_L \left(\frac{\sqrt{3}}{2} \cos \phi - \frac{1}{2} \sin \phi \right)
\end{aligned}$$

and

$$\begin{aligned}
P_2 &= V_L I_L \cos (\phi - \pi/6) \\
&= V_L I_L (\cos \phi \cos \pi/6 + \sin \phi \sin \pi/6) \\
&= V_L I_L \left(\frac{\sqrt{3}}{2} \cos \phi + \frac{1}{2} \sin \phi \right)
\end{aligned}$$

Note that ϕ is shown to exceed $\pi/6$ rad in the phasor diagram. If ϕ is less than $\pi/6$ the expression for P_1 remains unchanged both unexpanded and expanded, the expression for P_2 becomes

$$P_2 = V_L I_L \cos (\pi/6 - \phi)$$

which on expansion

$$= V_L I_L \left(\frac{\sqrt{3}}{2} \cos \phi + \frac{1}{2} \sin \phi\right)$$

as before.

The sum of the readings

$$P_1 + P_2 = \sqrt{3} \, V_L I_L \cos \phi$$

which is the total circuit power for a balanced system as shown earlier.

An expression for tan ϕ may be obtained as follows:

$$P_2 + P_1 = \sqrt{3} \, V_L I_L \cos \phi$$

and

$$P_2 - P_1 = V_L I_L \sin \phi$$

so that by division

$$\tan \phi = \frac{\sin \phi}{\cos \phi} = \sqrt{3} \left(\frac{P_2 - P_1}{P_2 + P_1}\right)$$

from which a value of ϕ and hence the power factor cos ϕ may be obtained by calculator or from tables.

An expression for cos ϕ may be derived using trigonometrical relationships as follows:

$$\text{since } \sec \phi = (1 + \tan^2 \phi)^{\frac{1}{2}}$$

$$\text{and } \sec \phi = \frac{1}{\cos \phi} \text{ then } \cos \phi = \frac{1}{\sec \phi} = \frac{1}{(1 + \tan^2 \phi)^{\frac{1}{2}}}$$

$$\cos \phi = \frac{1}{\left[1 + 3\left(\dfrac{P_2 - P_1}{P_2 + P_1}\right)^2\right]^{\frac{1}{2}}}$$

The two wattmeter method may be employed using a single instrument and appropriate switching or a special polyphase instrument containing two wattmeters in one housing.

Reactive volt-amperes   The reactive volt-amperes in a three-phase balanced load may be obtained from the readings of the two wattmeter method, for since reactive volt-amperes $= \sqrt{3} \, V_L I_L \sin \phi$ and the difference in wattmeter readings

$$P_2 - P_1 = V_L I_L \sin \phi$$

then

$$\text{reactive volt-amperes} = \sqrt{3} \, (P_2 - P_1)$$

Example 3.6   The two wattmeter method is used to measure the power absorbed by a three-phase induction motor running at 90% efficiency. The wattmeter readings are 1000 W and 12 050 W respectively. Calculate:

(a) the power absorbed by the machine;
(b) the load power factor and phase angle;
(c) the reactive volt-amperes taken by the load;
(d) the output power delivered by the motor.

(a) the power absorbed by the machine is the sum of the wattmeter readings, i.e.

$$12\,050 + 1000 = 13\,050 \text{ W}$$

The power absorbed is 13.05 kW.

(b) Denoting the wattmeter readings by P_1 and P_2 and taking $P_2 = 12\,050$ W and $P_1 = 1000$ W the power factor $\cos \phi$ is given by

$$\cos \phi = \frac{1}{\left[1 + 3 \left(\dfrac{P_2 - P_1}{P_2 + P_1}\right)^2\right]^{1/2}}$$

$$= \frac{1}{\left[1 + 3 \left(\dfrac{11\,050}{13\,050}\right)^2\right]^{1/2}}$$

$$= 0.563$$

The phase angle is arc cos 0.563 which, from tables or calculator, is equal to 0.972 rad.

The power factor is 0.563, the phase angle 0.972 rad.

(c) $\qquad$ Reactive volt amperes $= \sqrt{3}\,(P_2 - P_1)$
$$= \sqrt{3} \times 11\,050$$
$$= 19\,139 \text{ W}$$

Reactive volt amperes are 19.139 kVA.

(d) The power delivered to the machine is 13.05 kW and the machine efficiency is 0.9 so that

$$\text{Output power} = 0.9 \times 13.05 \text{ kW}$$
$$= 11.745 \text{ kW}$$

Output power is 11.745 kW.

Example 3.7 A balanced three-phase load absorbs 8.75 kW at a power factor of 0.7 lagging. Calculate the readings on each of two wattmeters connected to read the input power.

Denote the readings by P_1 and P_2

$$\text{total input power } P_1 + P_2 = 8750$$

which gives one relationship for P_1 and P_2. A second is required since here are two unknowns.

$$\text{Power factor } \cos \phi = \frac{1}{\left[1 + 3 \left(\dfrac{P_2 - P_1}{P_2 + P_1}\right)^2\right]^{1/2}}$$

Transposing

$$1 + 3 \left(\frac{P_2 - P_1}{P_2 + P_1} \right)^2 = \frac{1}{\cos^2 \phi}$$

and since $\cos \phi = 0.7$

$$1 + 3 \left(\frac{P_2 - P_1}{P_2 + P_1} \right)^2 = \frac{1}{0.7^2} = 2.04$$

so that

$$\left(\frac{P_2 - P_1}{P_2 + P_1} \right)^2 = \frac{2.04 - 1}{3} = 0.347$$

and

$$\frac{P_2 - P_1}{P_2 + P_1} = 0.347^{1/2} = 0.589$$

so that

$$P_2 - P_1 = 0.589 \, (P_2 + P_1)$$
$$= 0.589 \times 8750$$
$$= 5153.9$$

We now have

$$P_2 + P_1 = 8750$$
$$P_2 - P_1 = 5153.9$$

By addition

$$2P_2 = 13\,903.9$$
$$P_2 = 6951.9$$

and since

$$P_2 + P_1 = 8750$$
$$P_1 = 8750 - 6951.9$$
$$= 1798.1$$

The two readings are 1.8 kW and 6.95 kW, respectively.

It should be noted that in practice when using the two wattmeter method one of the readings may be negative and to read the meter the direction of indication may have to be reversed. The sign of the reading should always, however, be taken into account. Negative readings are obtained if the phase angle exceeds $\pi/3$ rad leading or lagging since under these conditions the current in one of the wattmeters has a component in antiphase to that of the voltage. This point is better understood by reference to fig. 3.9 which shows a phase angle of less than $\pi/3$ rad. The position of the current phasor for each wattmeter can be quite easily visualised if the phase angle exceeds $\pi/3$ rad.

Specific objectives   *The expected learning outcome is that the student:*
4.8   Draws phasor diagrams to identify the terms symmetrical and balanced.
4.9   Shows by phasor diagrams that the sum of line or phase currents in a balanced system is zero.

4.10 Draws phasor diagrams to explain normal and reverse phase rotation.

4.11 Calculates neutral current in a simple unbalanced 4-line system.

4.12 Shows by phasor diagrams that circulating voltage is twice line voltage in an incorrectly closed delta system.

Using phasor diagrams

As we have seen, problems involving three-phase systems may be solved by calculation using formulae and equations which have been previously derived. Phasor diagrams may also be used in problem solution and often providing an easier and shorter method. Phasor diagrams may also be used to illustrate certain basic characteristics of three phase systems, certain of which have already been derived mathematically.

Normal and reverse phase rotation

Earlier it was said that the three phases are denoted either by using numbers (1, 2, and 3) or colours (red, yellow and blue denoted R, Y and B respectively). 'Normal' phase rotation is RYB and a typical phasor diagram for line voltages might be as shown in fig. 3.10a.

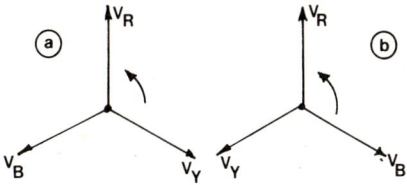

Figure 3.10

If the phase sequence were RBY the phasor diagram would be as shown in fig. 3.10b assuming anticlockwise rotation of phasors. This phasor diagram would also result from a phase sequence RYB but with the phasors rotating in the clockwise direction, i.e. in a *reverse* direction to that for 'normal' phase rotation. A phase sequence other than the normal RYB could then be referred to as 'reverse' rotation. It is extremely important to use the same rotation throughout a calculation, otherwise errors will be introduced. If the phase sequence or rotation is not given, 'normal' rotation can usually be assumed to be the correct one.

Symmetrical components

Fig. 3.11a shows a typical phasor diagram for the line currents of an unbalanced three-phase system with normal rotation. It can be

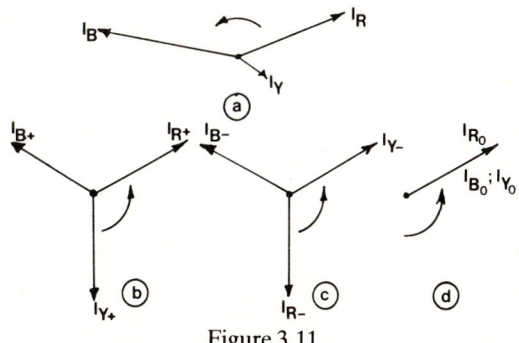

Figure 3.11

shown that this single phasor diagram would result from the superposition of three separate sets of phasors shown in fig. 3.11b, c and d respectively (superposition means placing all three diagrams on top of each other, having the same centre of rotation, and using phasor addition to produce a single phasor diagram with three phasors, one per current).

Fig. 3.11b shows a balanced system of currents with normal rotation, fig. 3.11c shows a balanced system of currents with reverse rotation. Figure. 3.11c shows three currents equal in phase and magnitude, there being no sequence associated with these currents.

The three phasors in fig. 3.11b are said to have *positive* phase sequence, those in fig. 3.11c to have *negative* phase sequence and those in fig. 3.11d to have *zero* phase sequence. These three sets of components of the phasors shown in fig. 3.11a are called *symmetrical components*. The phasor diagram of any unbalanced system may be broken down in this manner to give a set of symmetrical components, which may then be used as an additional aid to ease calculations and problem solution.

Sum of line or phase currents in a balanced system

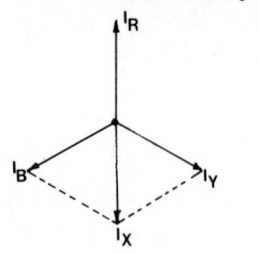

Figure 3.12

Phasor diagrams may be used to show that the sum of the line or phase currents in a balanced system is zero.

Fig. 3.12 shows a typical phasor diagram for the line currents of a three-phase balanced system of normal sequence and rotation. The resultant of currents I_Y and I_B, shown as I_X, is obtained by completing the phasor parallelogram of sides I_Y and I_B. If $I_Y = I_B$, I_x will bisect the angle between I_Y and I_B giving the angles shown and lie along the vertical. This may be verified by scale drawing which also shows I_X to equal I_Y or I_B. I_X is thus equal and opposite to I_R (since $I_R = I_Y = I_B$) and the resultant current is zero. This may also be shown using trigonometrical methods.

An incorrectly closed delta system

It was shown earlier that when a delta system is correctly closed, i.e. the direction of each of the phase voltages when positive is in the same direction round the loop, the resultant loop voltage is zero. If one of the phases is reversed the resultant loop voltage is not zero and is in fact equal to twice the line voltage. This may be shown using phasor diagrams as follows.

Fig. 3.13a shows three-phase coils having voltages V_R, V_Y and V_B having directions when positive as shown by the arrows. The phasor diagram for this arrangement is shown in fig. 3.13b. As can be seen the resultant of V_R and V_Y, shown as V_a is equal and opposite to V_B and the resultant loop voltage is zero. This was shown mathematically earlier in the chapter.

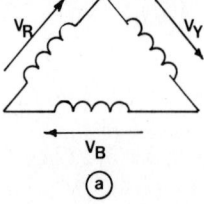

Figure 3.13

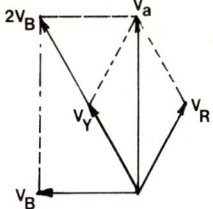

Figure 3.14

If now the phase coil of the yellow phase has its connections reversed, V_Y will act in the direction shown in fig. 3.14.

The resultant of V_R and V_Y, shown as V_a, is not now equal to V_R or V_Y (it is in fact equal to $\sqrt{3}\,V_R$ or $\sqrt{3}\,V_Y$) and does not lie in phase oposition to V_B. Using phasor addition to obtain the resultant of V_B and V_a, the resultant loop voltage is found to be *twice* the value of V_B (or V_R or V_Y), i.e. twice the line voltage. This may be easily seen by summing V_R and V_B to give a resultant lying on top of V_Y; the overall resultant is thus $2V_Y$ (or $2V_B$ or $2\,V_R$, i.e. $2 \times V_L$). Again this may be shown by applying standard methods of trigonometry to the phasor parallelograms shown in the figure.

Calculation of neutral current in a simple unbalanced 4-wire system

Phasor diagrams may also be conveniently used to calculate the neutral current in a simple unbalanced 4-wire system. For a balanced 4-wire system the current in the neutral line (the line connected to the star point) is zero. This is not the case when the system is unbalanced. See fig. 3.15.

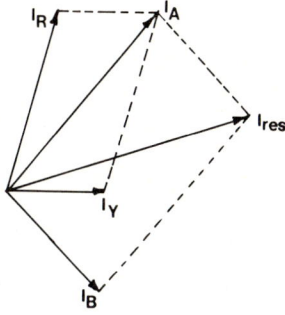

Figure 3.15

Fig. 3.15 shows a typical phasor diagram of an unbalanced three-phase system having line currents I_R, I_Y and I_B. Completion of the parallelogram gives I_A, the resultant of I_R and I_Y, and completion of the further parallelogram involving I_A and I_B gives the resultant of all the currents I_{res}, both in magnitude and direction. This current is the current which would flow in the neutral line of a three-phase 4-wire system in which the currents I_R, I_Y and I_B are the line currents.

Summary

A three-phase supply consists of separate voltages of equal amplitude and frequency separated in phase by $2\pi/3$ rad and generated simultaneously. Use is made of three phase in generation, transmission and distribution for reasons of efficiency, supply availability and improved machine operating characteristics.

The windings of a three-phase generator and the three arms of a three-phase load may be connected in star or delta. Star connection may have three or four supply lines depending upon whether or not the neutral point is used; delta connection always has only three lines. The resultant voltage in a correctly closed delta mesh (so that the phase voltages when positive act round the mesh in the same direction) is zero; if the mesh is incorrectly closed the resultant is equal to twice the phase voltage.

Denoting line voltage and current by V_L and I_L respectively and phase voltage and current by V_{ph} and I_{ph} respectively, for a star connected supply or load $V_L = \sqrt{3}\, V_{ph}$ and $I_L = I_{ph}$ and for a delta connected supply or load $V_L = V_{ph}$ and $I_L = \sqrt{3}\, I_{ph}$.

Power in a balanced three-phase load is $\sqrt{3}\, V_L I_L \cos \phi$ where $\cos \phi$ is the power factor, ϕ being the phase angle between V_L and I_L. Power may be measured by a single wattmeter (total power equals three times the reading if the system is balanced), by three wattmeters (total power equals sum of the readings, system balanced or otherwise) or by two wattmeters. The two wattmeter method is the most superior method and may be used on balanced or unbalanced systems. If the readings are P_1 and P_2, respectively, then

$$\text{total power} = P_1 + P_2$$

and the power factor

$$\cos \phi = \frac{1}{\left[1 + 3 \left(\dfrac{P_2 - P_1}{P_2 + P_1}\right)^2\right]^{1/2}}$$

The reactive volt amperes $= \sqrt{3}\,(P_2 - P_1)$.

Denoting phases by red, yellow and blue (R, Y, and B) normal phase rotation is R, Y, B. Normal rotation of phasors is anti-clockwise. The phasors of any unbalanced system may be replaced by three sets of balanced phasors, one rotating anticlockwise and containing three equal phasors, one rotating clockwise and containing three equal phasors and a single phase rotating anticlockwise. These three sets of phasors are called symmetrical components of the phasors representing the unbalanced voltages or currents. Symmetrical components have a positive phase sequence (anticlockwise), negative phase sequence (clockwise) or zero phase sequence (anticlockwise).

EXERCISE 3.1

1. Three equal impedances are connected in delta to a 440 V, 50 Hz supply. The power factor is 0.8 lagging and 25 kVA is drawn from the supply. Calculate the line current and total power drawn from the same supply when the same impedances are star connected.

2. A three-phase star connected generator supplies 230 V per phase to a delta connected load composed of equal impedances of resistance 50 Ω and inductive reactance 120 Ω. Calculate the:
    (a) load phase current;
    (b) load power factor;
    (c) the line current;
    (d) generator output.

3. A 440 V three-phase delta connected motor has an output of 14 kW with an efficiency of 92% and power factor of 0.9. Calculate the line current.

4. Calculate the line and phase currents in a balanced delta connected load taking 75 kW at a power factor of 0.8 from a three-phase 440 V supply.

5. Calculate the total power supplied to a star connected resistive load of 150 Ω per phase when the line voltage is 240 V.

6. A three-phase 440 V, 50 Hz motor on full-load has an efficiency of 90%, a power output of 15 kW and a power factor of 0.82 lagging. Calculate the input kW, kVA and kVAr and determine the full load phase currents assuming delta connection.

7. A balanced three-phase load takes 10 kW at a power factor of 0.9 lagging. Calculate the readings on each of two wattmeters connected to read the input power.

8. The readings on the meters in the two wattmeter method for determining three-phase power applied to a certain load were 7.5 kW and 8.2 kW. Calculate the total power, reactive volt-amperes and power factor.

Possible
marks

SELF-ASSESSMENT EXERCISE 3   State the relationship between line and phase voltages and currents for a star connected three-phase system. (3)

2. State the relationship between line and phase voltages and currents for a delta connected three-phase system. (3)

3. State the equation for the total power absorbed by a three-phase load in terms of line voltage, line current and power factor. (3)

4. Three wattmeters connected to read the power in a three-phase load read 4.2 kW, 7.8 kW and 9.6 kW respectively. Determine the total power absorbed by the load. (3)

5. Calculate the phase voltage for the following line voltages for the connection shown:
   (a) 440 V star;
   (b) 240 V delta. (3)

6. State the equation relating the total power absorbed, the reactive volt-amperes and the power factor of a three-phase load in terms of the readings P_1 and P_2 on two wattmeters. (5)

7. In the two wattmeter method for measuring three-phase power the readings were 8.6 kW and 9.2 kW. Calculate the total power, the reactive volt-amperes and the power factor. (5)

8. State three advantages of using three-phase in electricity supply generation, distribution and transmission. (5)

9. Show by phasor diagrams drawn to scale the relationship between line and phase voltages in a balanced three-phase star connected system.

A balanced delta connected load takes 50 kW from a three-phase 440 V supply at a power factor of 0.9 lagging. Calculate the line and phase currents and the phase voltage of the system. (14)

10. State the equation relating the power supplied to a balanced three-phase system in terms of line voltage, line current and power factor. Calculate the total power supplied to a delta connected resistive load of 100 Ω/phase by a three-phase generator having a line voltage of 230 V. (14)

11. Three identical inductive loads of resistance 15 Ω and reactance 40 Ω are connected in star to a 440 V, three-phase supply. Calculate the line and phase currents and the power absorbed from the supply. (14)

12. State the principal advantage of using two wattmeters to measure power in a three-phase load.

The total power absorbed by a cetain three-phase load is 60 kW the reactive volt-amperes being 100 kVAr. Determine the power factor of the load and the readings on two wattmeters connected to read the load power. (14)

13. With the aid of phasor diagrams:
(a) show that the resultant voltage in an incorrectly closed delta connected three-phase system is twice the phase voltage. What is meant by 'incorrectly closed'?
(b) show that the resultant current in a balanced three-phase star connected load is zero.
(c) explain what is meant by 'symmetrical components'.   . (14)

Answers

EXERCISE 3.1

1. 18.94 A; 11 547 W

2. (a) 1.77 A (b) 0.38 (c) 3.066 A (d) 803.9 W

3. 22.19 A

4. 123 A line, 71.02 phase

5. 384 W

6. 16.67 kW; 20.32 kVA; 11.63 kVAr; 15.4 A

7. 6398 W; 3602 W

8. 15.7 kW; 1.21 kVAr; 0.997

		Marks
SELF-ASSESSMENT EXERCISE 3	1. $V_L = \sqrt{3}\, V_{ph}$	(1½)
	$I_L = I_{ph}$	(1½)
	2. $V_L = V_{ph}$	(1½)
	$I_L = \sqrt{3}\, I_{ph}$	(1½)
	3. $\sqrt{3}\, V_L I_L \cos\phi$	(3)

4.   Total power = sum of wattmeter readings   (1)
= (4.2 + 7.8 + 9.6) kW   (1)
= 21.6 kW   (1)

5. (a)   Phase voltage = $440/\sqrt{3}$
= 254 V   (1½)

   (b)   phase voltage = line voltage
= 240 V   (1½)

6.   Total power = $P_2 + P_1$ (1½)
Reactive volt-amperes = $\sqrt{3}\,(P_2 - P_1)$ (1½)

Power factor = $\dfrac{1}{\left[1 + 3\left(\dfrac{P_2 - P_1}{P_2 + P_1}\right)^2\right]^{1/2}}$ (2)

7. $P_2 = 9.2$ kW $P_1 = 8.6$ kW

Total power = $P_2 + P_1 = 17.8$ kW   (1½)

Reactive volt amperes = $\sqrt{3}\,(P_2 - P_1)$
= 1.04 kVAr   (1½)

Power factor = cos [arc tan (1.04/17.8)]
= 0.998   (2)

(alternatively the equation given in the chapter can be used).

8. (a) Higher efficiency   (1½)
(b) Availability of industrial (three-phase) and domestic (single-phase) supplies from same source.   (1½)

(c) Superior characteristics of three-phase machines including higher efficiency and power factor.   (2)

9. Phasor diagrams as text (fig. 3.6).   (5)

$$\text{Power} = 50\,000 \text{ W}$$
$$\text{Therefore } \sqrt{3}\, V_L I_L \cos\phi = 50\,000$$
$$\sqrt{3} \times 440 \times I_L \times 0.9 = 50\,000$$

and $I_L = \dfrac{50\,000}{\sqrt{3} \times 440 \times 0.9} = 72.89$ A   (3)

$$I_{ph} = \frac{I_L}{\sqrt{3}} = 42.08 \text{ A} \quad (3)$$

$$V_{ph} = V_L$$
$$= 440 \text{ V} \quad (3)$$

10. Equation is $\sqrt{3}\, V_L I_L \cos\phi$ (1)

$$V_{ph} = V_L = 230 \text{ V} \quad (3)$$

$$I_{ph} = \frac{230}{100} = 2.3 \text{ A} \quad (3)$$

$$I_L = \sqrt{3}\, I_{ph} = 3.98 \text{ A} \quad (3)$$

$$\cos\phi = 1 \text{ (resistive load)} \quad (3)$$

$$\text{power supplied} = \sqrt{3} \times 230 \times 3.98 \times 1$$
$$= 1585.5 \text{ W} \quad (1)$$

11. $\text{Phase impedance} = (15^2 + 40^2)^{1/2}$
$$= 42.72 \ \Omega \quad (2)$$

$$V_{ph} = \frac{V_L}{\sqrt{3}} \text{ (star connection)} \quad (2)$$

$$= 440/\sqrt{3} = 254 \text{ V} \quad (1)$$

Therefore $I_{ph} = V_{ph}/\text{phase impedance}$ (2)
$$= 254/42.72$$
$$= 5.94 \text{ A} \quad (1)$$
$$I_L = I_{ph} = 5.94 \text{ A} \quad (2)$$

Power factor = $\cos(\arctan 40/15) = 0.35$ (2)

So power = $\sqrt{3}\, V_L I_L \cos\phi$ (1)
$$= \sqrt{3} \times 440 \times 5.94 \times 0.35$$
$$= 1589.5 \text{ W} \quad (1)$$

12. The principal advantage is that the method may be used regardless of the state of balance, waveform, phase sequence or load connection.   (3)

Taking the wattmeter readings as P_2 and P_1

power absorbed = $P_2 + P_1$ (2)
reactive volt-amperes = $\sqrt{3}\,(P_2 - P_1)$ (2)

so that $60\,000 = P_2 + P_1$
$100\,000 = \sqrt{3}\,(P_2 - P_1)$

i.e. $\dfrac{100\,000}{\sqrt{3}} = P_2 - P_1$

$$57\,735 = P_2 - P_1 \quad (1)$$
$$\text{and } 60\,000 = P_2 + P_1 \quad (1)$$

By addition $2P_2 = 117\,735$
and $P_2 = 58\,867.5$ (1)
By subtraction $2P_1 = 2265$
and $P_1 = 1132.5$ (1)

Power factor $= \cos\,[\text{arc tan (kVAr/kW)}]$ (2)
$\qquad\qquad\quad = \cos\,(\text{arc tan } 100/60)$
$\qquad\qquad\quad = 0.51$ (1)

(Or the alternative equation relating $\cos\phi$, P_2 and P_1 may be used).

13. (a) Phasor diagram as text (fig. 3.13) plus explanation definition of 'incorrectly closed' – see text. (4) (1)

(b) Phasor diagram as text (fig. 3.12) plus explanation. (4)

(c) Phasor diagrams as text (fig. 3.11) plus explanation (to include mention of phase sequence). (5)

4 D.C. Transients

Topic area: D

General objective

The expected learning outcome is that the student understands transient behaviour of simple C-R circuits.

As was shown in *Electrical and Electronic Principles 2*, when a component has capacitance it opposes *changing* voltage; when a component has inductance it opposes *changing* current. A capacitive component does not oppose a voltage of constant value and, similarly, an inductive component does not oppose a current of constant value. A resistive component on the other hand offers the same opposition to current whether it is changing or constant in value. The important word when considering the effects of capacitive and inductive components is *changing* because different effects are observed in circuits when voltages and currents are changing and when they are not. In d.c. circuits in which currents and voltages are not usually continually changing the effects of capacitance and inductance are noticeable only on first connecting the supply voltage to the circuit or disconnecting it from the circuit. In a.c. circuits in which, as we have seen, currents and voltages are continually changing, the effects of capacitance and inductance are present all the time although even here the effects are more pronounced when the supply voltage is first connected to or disconnected from the circuit than when the circuit has been active for some time. This period during which these effects are more pronounced is called the *transient* period, the word transient meaning 'temporary' or 'passing'. During the transient period additional voltages (back e.m.f.s) appear in inductive circuits and additional (charging) currents appear in capacitive circuits and these additional voltages and currents (which disappear when the transient period is over) are called transient voltages and currents, or, merely transients. When all the transients have disappeared the circuit is then said to be in the *steady state*. In this chapter we shall only be examining the transients which occur in d.c. circuits although it should be remembered that even in a.c. circuits where change is always occurring the magnitudes and waveforms of currents and voltages are different in the transient period than in the steady state period.

Specific objectives

The expected learning outcome is that the student:
5.1 Explains how the current and capacitor voltage in a series C-R circuit which is connected to a d.c. source, vary with time.
5.2 Sketches the curves for the variation of voltage and current with

time for each of the components in a series C-R circuit when the
capacitor is:
(a) charging;
(b) discharging.
5.3   Defines the time-constant of a series C-R circuit.
5.4   Determines the growth and decay of the component voltage or
      current in a series C-R circuit, seconds after the commencement
      of:
      (a) charging;
      (b) discharging.

C-R circuits

As we have seen, capacitance is the characteristic of storing electric
charge. Components especially designed to have the characteristic
are called capacitors but it should always be borne in mind that
other components or parts of circuits, for example, connecting
leads, may also have capacitance to a greater or lesser degree and
this will also have an effect on what occurs during the transient
period. In the discussion which follows the word capacitor will be
used for convenience; in a practical circuit there may be an actual
capacitor or an equivalent capacitor, i.e. one having the capacitance
of the circuit and used for calculation purposes, or, of course, a
mixture of both.

When a voltage is applied to a capacitor a transient current flows
in the connecting leads. The current flow deposits electrons on one
plate (or set of plates) of the capacitor, causing it to become
negatively charged, and removes electrons from the other plate,
causing it to become positively charged. Current does not flow
between the plates, the action of one plate on the other being due to
the electric field set up between the plates.

As the one plate becomes increasingly negative the repulsion
between it and the electrons moving onto the plate increases and the
process of charging slows down and eventually stops. At this
juncture, the steady state, the capacitor is said to be fully charged,
no current flows in the circuit and the voltage present is that across
the capacitor.

If a conductive lead is now connected across the capacitor the
additional electrons on the negatively charged plate move through
the lead to the positive plate and discharge commences. Again a
transient current flows until the capacitor is completely discharged.
During the discharge period the electric field in the capacitor
dielectric collapses and the voltage across the capacitor decays to
zero. In the new steady state there is no voltage or current in the
circuit.

Typical graphs of current and voltage plotted against time during
which a capacitor is charged or discharged are shown in fig. 4.1a and
b the circuit to which the graph applies being shown in fig. 4.1c.

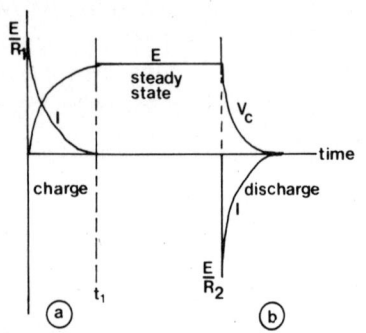

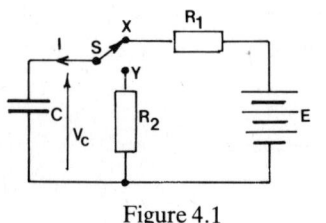

Figure 4.1

Charging

The time axis of the graph begins (at zero) at the instant of placing
switch S in position X shown in fig. 4.1c. When the switch is closed a
charging current *I* begins to flow in the circuit. At the instant of

closure there is no voltage across the capacitor (since this cannot change instantaneously) and the whole of the applied voltage E appears across the resistor. The circuit current I is thus equal to E/R_1 and the capacitor voltage $V_C = 0$.

The charging process now begins and V_C rises in value as the charging current falls in value as shown in fig. 4.1a. As explained earlier, as charge accumulates on the capacitor plates the *rate of charge* slows down and this is shown in the graph by the slope of the V_C/time curve becoming progressively *less* as time passes. The shape of the current/time graph is similar, as it is the current which is doing the charging, but on this occasion the current is reducing in value as the capacitor voltage is increasing.

This type of curve in which the rate of change of a variable quantity depends upon the value of the quantity (as here the rate of change of current I or voltage V_C at any particular instant depends upon the value of current I or voltage V_C at that instant) is well known in mathematics. It is called an *exponential* curve and the mathematical equation describing the curve contains the constant (exponent) e. The constant e is one of several such constants, another example being π, which occur fairly regularly in mathematical relationships. The value of the constant e is 2.7183 to four decimal places.

$$V_C = E\left(1 - e^{-t/CR_1}\right)$$

and
$$I = \frac{E}{R_1}e^{-t/CR_1}$$

which at first sight may seem a little complicated. A further examination of these equations will help to clarify them.

$$e^{-t/CR_1} \text{ means } 1/e^{t/CR_1}$$

which, in turn, means the reciprocal of e raised to the power t/CR_1.

When

$$t = 0, \frac{t}{CR_1} = 0 \text{ and } e^{t/CR_1} = e^0 = 1$$

(since any variable raised to the power zero is equal to 1)

so that, when $t = 0$, $V_C = E(1 - 1)$, i.e. $V_C = 0$ and

$$I = \frac{E}{R_1}$$

and these values are shown on the graph.

At a time shown as t_1 on the graph the steady state period begins. At this point the capacitor voltage has reached the value of E and the charging current I has fallen to zero (later we shall be looking a little more closely at how long it takes to reach the steady state from the instant of the start of charging).

Discharging

The steady state period shown in fig. 4.1a lasts until the switch position is changed from X to Y. With the switch in position Y the capacitor is placed in parallel with resistor R_2 and will begin to discharge as shown in fig. 4.1b. At the instant of switching the capacitor voltage, which has the value E volts since there has been a steady state period, is across the resistor R_2 and the initial discharge current I has the value E/R_2 amperes. This current flows in the *opposite* direction to that in which the charging current flowed so it is shown in the opposite or negative direction on the graph. As the current flows the capacitor loses its charge, i.e. the excess electrons on the negative plate move through resistor R_2 to the positive plate and since the capacitor voltage

$$V_C = Q/C$$

where Q is the capacitor charge, V_C starts to decay. Since the discharge current I has the value given by the equation

$$I = V_C/R_2$$

it too begins to fall from its initial value of E/R_2 as shown in the graph. (Since I is shown as negative it can be said that since it is becoming *less* negative, i.e. *more* positive, its value is *rising* but on this occasion it is less confusing to consider values without polarity in which case the value is falling.)

Again the change in both capacitor voltage and current is an exponential change and the equations describing the change contain the constant e.

On discharge

$$V_C = E\,e^{-t/CR_2}$$

and

$$I = \frac{-E}{R_2}e^{-t/CR_2}$$

where t represents the value of time measured after the switch is moved to position Y, i.e. prior to $t = 0$ the condition is the steady state.

When $t = 0$,

$$V_C = E$$

and

$$I = \frac{-E}{R_2} \text{ (since e}^0 = 1)$$

Time constant

The values of capacitance and resistance in the circuit determine the time taken to completely charge or discharge the capacitor. If the capacitance is high it will acquire more charge per volt (since $Q = CV$) and for a given level of current will therefore take longer to charge (current, remember, is charge per unit time). Similarly, if the resistance is high the charging current, which starts at a value E/R and then falls, will be lower for a given value of E and again the

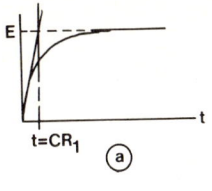

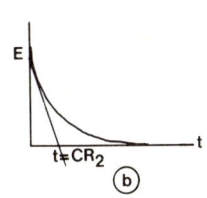

Figure 4.2

charging process takes longer. The same comments apply to discharge, for here again the larger the capacitance the greater is the initial level of charge and the larger the resistance the smaller is the initial and subsequent level of discharge current. Clearly both charge and discharge times are affected by the values of capacitance or resistance.

The *rate* of charge or discharge at any particular time is shown by the gradient or slope of the V_C/time graph at that time. A tangent drawn on the graph at any point indicates the slope and thus the rate of charge or discharge.

In fig. 4.2a a tangent is drawn on the graph at the point of switching and shows the rate of charge at this point; similarly the tangent in fig. 4.2b shows the rate of discharge at the instant of switching.

If the rate of charge or discharge were not to change but were to remain at the same rate as that at the instant of switching it is found that the capacitor would be completely charged or discharged in a time (in seconds) equal to the product of the capacitance (farads) and the appropriate resistance (ohms), i.e. R_1 when charging and R_2 when discharging. This is shown on the graph by continuing the tangent in fig. 4.2a to the voltage level E line and in fig. 4.2b to the zero voltage line.

This product capacitance × resistance is an important parameter in a capacitive circuit and is called the *time constant* of the circuit, symbol τ (pronounced 'tor'). As was shown earlier the values of C and R affect the time of charge or discharge and this can now be seen, for, the larger the values of C and R, the greater is the value of the time constant CR. As CR is increased the shallower is the slope of the initial tangent, i.e. the initial rate of charge or discharge is smaller and the longer it is going to take to complete the charging or discharging process. We shall look again shortly at how long it actually takes for the process to be complete. The fact that if the initial rate of charge or discharge were to remain the same it would take CR seconds for the process to be complete can be shown mathematically as follows:

When charging

$$V_C = E\,(1 - e^{-t/CR_1})$$

The rate of change of V_C with time

$$\frac{\mathrm{d}V_C}{\mathrm{d}t} = \frac{\mathrm{d}}{\mathrm{d}t}\,(E - E e^{-t/CR_1})$$

$$= \frac{E}{CR_1}\,e^{-t/CR_1}$$

and when $t = 0$,

$$\frac{\mathrm{d}V_C}{\mathrm{d}t} = \frac{E}{CR_1}$$

i.e. V_C is charging by E volts every CR_1 seconds so that after CR_1 seconds V_C has changed from zero (at $t = 0$) to E.

Similarly on discharge

$$V_C = E\,e^{-t/CR_2}$$

$$\frac{dV_C}{dt} = -\frac{E}{CR_2}\,e^{-t/CR_2}$$

and when $t = 0$,

$$\frac{dV_C}{dt} = -\frac{E}{CR_2}$$

the same form of equation as on charge but a negative rate of change, i.e. V_C is falling.

The value of the capacitor voltage after a time equal to the time constant has elapsed after the instant of switching may be obtained by putting $t = CR_1$ or CR_2 in the appropriate equation.

$$\begin{aligned}
\textit{Charging } V_C &= E\,(1 - e^{-CR_1/CR_1})\\
&= E\,(1 - 0.368) \text{ from tables}\\
&= 0.632E
\end{aligned}$$

$$\begin{aligned}
\textit{Discharging } V_C &= E\,e^{-CR_2/CR_2}\\
&= E\,e^{-1}\\
&= 0.368\,E
\end{aligned}$$

and we see that in a time equal to the time constant the capacitor voltage when charging has *changed* by 0.632 of the final voltage E or when discharging has changed by 0.632 of the initial voltage E. We can then define the time constant as the time taken for the capacitor voltage to change by 0.632 or 63.2% of its final or initial value depending upon whether the capacitor is being charged or discharged.

When is the charging or discharging process completed? By inserting values of time equal to multiples of the time constant, i.e. CR, $2CR$, $3CR$, $4CR$ etc. we can see how long it takes to complete the process

Time	Charging	Discharging
CR	$V_C = 0.632E$	$V_C = 0.368E$
$2CR$	$V_C = 0.865E$	$V_C = 0.135E$
$3CR$	$V_C = 0.95E$	$V_C = 0.049E$
$4CR$	$V_C = 0.982E$	$V_C = 0.018E$
$5CR$	$V_C = 0.993E$	$V_C = 0.007E$

and it appears that the process is never actually completed. This is a characteristic of an exponential curve; it moves closer to a final value but never gets there! However, for all practical purposes we can take $5CR$ as a reasonable time for completion for taking the constant preceding E to one decimal place, after $5CR$ seconds

$$V_C = E \text{ (charging)}, \quad V_C = 0 \text{ (discharging)}$$

Example 4.1   A 0.5 μF capacitor is connected to a 200 V supply via a 150 Ω resistor. Ignoring lead resistance, calculate the circuit time constant and the capacitor voltage after a time equal to the time constant.

$$\text{Time constant} = CR \text{ seconds}$$
$$= 0.5 \times 10^{-6} \times 150$$
$$= 75 \ \mu\text{s}$$

After a time equal to the time constant

$$V_C = 0.632E$$
$$= 0.632 \times 200$$
$$= 126.4 \text{ V}$$

Example 4.2   A 10 μF capacitor is fully charged via a total resistance of 22 kΩ to 250 V. Calculate the capacitor voltage 10 ms after charging commenced. How long did it take for the capacitor to be fully charged?

During charge

$$V_C = E \left(1 - e^{-t/CR}\right)$$

so that

$$V_C = 250 \left(1 - e^{-10 \times 10^{-3}/10 \times 10^{-6} \times 22 \times 10^3}\right)$$
$$= 250 \left(1 - e^{-0.045}\right)$$
$$= 250 \left(1 - 0.955\right)$$
$$= 11 \text{ V}$$

The circuit time constant is $10 \times 10^{-6} \times 22 \times 10^3$ s which equals 0.22 s. We can assume that the capacitor was fully charged, therefore, after 5×0.22, i.e. 1.1 s.

(After 1.1s the capacitor voltage is actually given by

$$V_C = 250 \left(1 - e^{-5}\right)$$
$$= 248.31 \text{ V}$$

This calculation would not usually be expected in such an answer).

Example 4.3   A capacitor is fully charged to a p.d. of 200 V. When discharged through a 250 Ω resistor the capacitor voltage falls to 45 V in 0.3 s. Calculate the capacitance of the capacitor.
The equation for the capacitor voltage on discharge, using the usual symbols, is

$$V_C = E \, e^{-t/CR}$$

so that, here,

$$45 = 200 \, e^{-0.3/CR}$$

Hence

$$e^{-0.3/CR} = \frac{45}{200}$$

and

$$e^{0.3/CR} = \frac{200}{45} = 4.44$$

$$\frac{0.3}{CR} = \ln 4.44 \text{ (using natural logarithms)}$$

$$CR = 0.3/\ln 4.44$$
$$= 0.2$$

$$\text{and } C = \frac{0.2}{R}$$

$$= \frac{0.2}{250}$$

$$= 8.04.5 \times 10^{-6} \text{ F}$$
$$= 804.5 \ \mu\text{F}$$

This capacitance is $804.5 \ \mu\text{F}$.

Notice that natural logarithms are used here. Natural logarithms are those to the base e and can be obtained from tables or by use of a scientific calculator. The latter is recommended.

Summary The voltage across a capacitor cannot change instantaneously. When a voltage is connected across a capacitor the capacitor voltage rises exponentially and its value V_C at any time t seconds after connection is given by

$$V_C = E \left(1 - e^{-t/CR}\right) \text{ volts}$$

where

E is the applied voltage (volts)
C is the capacitance (farads)
R is the *total* resistance through which the charging current flows (ohms)

The charging current rises instantaneously on connection to a value given by E/R amperes and then decays exponentially to zero. The value of the charging current I at any time t seconds after connection is given by

$$I = \frac{E}{R} e^{-t/CR} \text{ amperes}$$

When a capacitor is discharged the capacitor voltage decays exponentially, the voltage V_C any time t seconds after discharge has begun being given by

$$V_C = E \, e^{-t/CR} \text{ volts}$$

where

E is the initial value of capacitor voltage when discharge begins (volts)
C is the capacitance (farads)

R is the total resistance through which the discharge current flows (ohms)

The discharge current rises instantaneously to a value of *E/R* amperes (flowing in the opposite direction to that in which the charging current flows) and then decays exponentially to zero. The discharge current *I* any time *t* seconds after discharge has begun is given by

$$I = \frac{E}{R}\,\mathrm{e}^{-t/CR} \text{ amperes}$$

In these four equations the quantity obtained by multiplying together circuit capacitance, *C* farads, and circuit resistance, *R* ohms, i.e. *CR* seconds, is an important parameter of the circuit and is called the time constant. It is the time in which the change in capacitor voltage would be complete if the initial rate of change were to be maintained. The initial rate of change does not remain constant, since the change is exponential, and after a time equal to the time constant the capacitor voltage in fact has changed by 0.632 of the final value if the capacitor is charging or 0.632 of the initial value if the capacitor is discharging.

Specific objectives

The expected learning outcome is that the student:

5.5 *Explains the growth and decay of current and voltages in a series L-R circuit.*

5.6 *Sketches curves for the variation of voltage and current with time for each of the components in a series L-R circuit after the circuit has been*
   (a) connected to, and
   (b) disconnected from a d.c. supply.

5.7 *Defines the time constant of a series L-R circuit.*

5.8 *Calculates the component voltage or current in a series L-R circuit, seconds after the circuit has been:*
   (a) connected to, and
   (b) disconnected from, a d.c. supply.

L-R circuits

Inductance, as has been shown in an earlier unit, is the property of opposing changing current. The opposition is established by a changing magnetic field which in turn sets up a changing voltage called a back e.m.f. This back e.m.f. acts so as to oppose the current change, attempting to reduce the current level if it is rising and to maintain the current level if it is falling. As with *C-R* circuits the nature of the changes, in an inductive circuit on this occasion of the current and of the back e.m.f., is exponential. This is shown in fig. 4.3. Fig. 4.3a shows graphs of the current and back e.m.f. plotted against time elapsed after an inductive circuit has been closed. The circuit to which the graphs apply is shown in fig. 4.3b.

On closing switch S in the circuit shown the whole of the battery voltage is applied to the inductive coil (neglecting any voltage occurring at the battery terminals due to its own internal resistance)

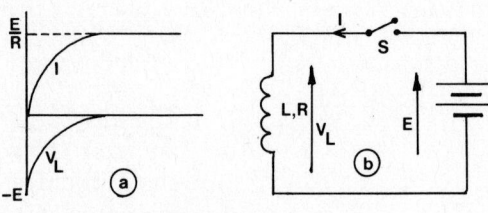

Figure 4.3

but the circuit current cannot rise immediately to the final value E/R amperes because of the back e.m.f. established by the inductive nature of the coil. At the instant of switching this back e.m.f., shown as V_L, is equal in value, and, of course, acts in the opposite direction, to the battery voltage. The value of the back e.m.f. at any instant is directly dependent upon the rate of change of circuit current, since, as was shown in an earlier unit,

$$E = -L\frac{di}{dt}$$

and at the instant of switching the rate of change of circuit current is at a maximum, as shown in fig. 4.3a.

The current now rises exponentially, its rate of change (indicated by the *slope* of the current/time graph) becoming less as the time after switching increases, so that the decay of the back e.m.f. also follows an exponential curve as shown.

The equations connecting circuit current, I, and back e.m.f., V_L, with the remaining circuit constants are as follows:

$$I = \frac{E}{R}(1 - e^{-Rt/L})$$

and

$$V_L = E\,e^{-Rt/L}$$

where R is the *total* circuit resistance (ohms), including in a practical circuit the resistance of all connecting leads between voltage source and coil, and L is the inductance of the coil (henrys).

Time constant   When $t = L/R$

$$I = \frac{E}{R}(1 - e^{-1})$$

$$= 0.632\,E/R$$

$$\text{and } V_L = E\,e^{-1}$$
$$= 0.378\,E$$

i.e. the current is equal to and has therefore *changed by* 0.632 of its maximum value and the back e.m.f. has also changed by 0.632 of its *initial* value. The parameter L/R is thus the circuit time constant τ and, as might be expected, if a tangent is drawn to the current/time curve at the instant of switching and continued to the maximum current line the time taken to reach it is found to be equal to L/R seconds. See fig. 4.4.

Figure 4.4

The figure shows that if it were possible to maintain the rate of change of current at the value it has at the instant of switching, the time taken to complete the change, i.e. to reach the maximum level of current would be L/R seconds.

Examining the factors which determine the time constant in this case we see that time constant is directly proportional to circuit inductance L and indirectly proportional to circuit resistance R.

The physical reasons for this are as follows. The circuit inductance determines the *opposition* to change (the back e.m.f. is equal to inductance $\times$ rate of change of current) so that the larger the inductance the longer it takes for the circuit current to reach the steady state value, and thus the greater the value of the time constant. The circuit resistance, of course, helps determine the steady state value; the greater it is the smaller the value of the steady state current so that the greater the resistance, the smaller the value of the final current and the less time it takes to reach it, i.e. the smaller the value of the time constant.

Opening inductive circuits poses problems. When switch S is opened and the inductor magnetic field begins to collapse a back e.m.f. is induced which attempts to maintain current flow and the current does not therefore immediately fall to zero. The result in a highly inductive circuit is a spark occurring at the switch as current flows in the gap between contacts and the contacts become increasingly damaged as the switch is used again and again. Switches for highly inductive circuits are often especially designed so that the break between contacts is made extremely quickly, thus rapidly increasing circuit resistance and reducing the time constant or, alternatively, a separate resistor is switched into the circuit slightly before the switch between inductor and supply is opened so that the decaying current flows in the resistor and does not attempt to jump the contact gap.

On opening the circuit the inductor current I decays exponentially, the relationship between it and the other circuit quantities being

$$I = \frac{E}{R}\,\mathrm{e}^{-Rt/L}$$

In both these equations R is the *total* circuit resistance, i.e. the resistance of the inductor *and* that of the gap between contacts if the switch is opened without inserting a separate resistance as it does so. If a separate resistance is inserted to prevent sparking then the total circuit resistance is the sum of this resistance and the inductor resistance.

As before the time constant, L/R seconds, is the time taken for either I or V_L to change by 0.632 of their original value and would be the time taken to reach the steady state if the original rate of change of I or V_L were to be maintained.

Example 4.4 A coil of inductance 10 H and resistance 200 Ω is connected to a 50 V supply. Calculate the value of the current in the circuit 20 ms after switching and determine the time constant of the circuit.

Using the relationship

$$I = \frac{E}{R}(1 - e^{-Rt/L})$$

after 20 ms, since $E = 50$ V, $R = 200$ Ω and $L = 10$ H

$$I = \frac{50}{200}(1 - e^{-200 \times 20 \times 10^{-3}/10})$$

$$= 0.25(1 - e^{-0.4})$$
$$= 0.082 \text{ A}$$

After 20ms the current is 82 mA.

The circuit time constant is L/R seconds, i.e. 10/200 s or 50 ms.

Ex 4.5

Example 4.5   The steady state current flowing in an inductor is 250 mA; the current flowing 0.1 s after connecting the supply voltage is 120 mA. Calculate the circuit time constant and the time from closing the circuit at which the circuit current has reached 200 mA.

The current I at any time t seconds after connecting the supply voltage is given by

$$I = I_s\left(1 - e^{-t/\tau}\right)$$

mistake   ✳

$$I = I_s(1 - t/\tau) \text{ amperes}$$

where I_s is the steady state current (amperes) and τ is the time constant (seconds). So that, since in this example,

$$I_s = 250 \text{ mA and } I = 120 \text{ mA when } t = 0.1 \text{ s}$$

$$120 \times 10^{-3} = 250 \times 10^{-3}(1 - e^{-0.1/\tau})$$

hence $1 - e^{-0.1/\tau} = \dfrac{120}{250} = 0.48$

and $e^{-0.1/\tau} = 1 - 0.48 = 0.52$

$e^{0.1/\tau} = 1/0.52 = 1.923$

Using natural logarithms:

$$0.1/\tau = \ln 1.923$$
$$= 0.6539$$
$$\text{and } \tau = 0.1/0.6539 = 0.1535$$

The time constant is 0.153 s.

When the circuit current is 200 mA,

$$200 \times 10^{-3} = 250 \times 10^{-3}(1 - e^{-t/0.153})$$

$$\text{and } 1 - e^{-t/0.153} = \frac{200}{250} = 0.8$$

$$e^{-t/0.153} = 0.2$$
$$e^{t/0.153} = 1/0.2 = 5$$

$$\frac{t}{0.153} = \ln 5 = 1.609$$

and
$$t = 1.609 \times 0.153$$
$$= 0.25 \text{ s}$$

The circuit current is 200 mA after 0.25 s has elapsed from closing the circuit.

Example 4.6   An inductive coil of inductance 1 H and resistance 100 Ω is supplied from a 50 V d.c. source of internal resistance 5 Ω. Calculate:

   (a) the circuit time constant
   (b) the steady state current
   (c) the time elapsed from closing the circuit at which the circuit current has reached 35% of the steady state value.

(a) The circuit time constant is given by

$$\frac{\text{inductance}}{\text{total circuit resistance}} = \frac{1}{105} \text{ s}$$

(total circuit resistance = coil resistance + supply internal resistance)

$$= 9.52 \text{ ms}$$

(b) The steady state current is

$$\frac{\text{supply voltage}}{\text{total circuit resistance}}$$

$$= \frac{50}{105}$$

$$= 0.476 \text{ A}$$

(c) When the current has reached 35% of the steady state current it is equal 0.35×0.476 A, i.e. 0.1667 A. So that

$$0.1667 = 0.476 \left(1 - e^{-t/9.52 \times 10^{-3}}\right)$$

and
$$1 - e^{t/9.52 \times 10^{-3}} = \frac{0.1667}{0.476} = 0.35$$

$$e^{-t/9.52 \times 10^{-3}} = 1 - 0.35 = 0.65$$

$$e^{t/9.52 \times 10^{-3}} = \frac{1}{0.65} = 1.54$$

$$\frac{t}{9.52 \times 10^{-3}} = \ln 1.54 \times 0.43$$

$$t = 0.43 \times 9.52 \times 10^{-3}$$
$$= 4.1 \text{ ms}$$

Units of time constant

Time constant τ in either *L-R* or *C-R* circuits is the time after which the changing quantity has changed by 0.632 of its initial value and is measured in seconds.

In a *C-R* circuit $\tau = CR$
In an *L-R* circuit $\tau = L/R$

The unit of capacitance is the farad, of resistance is the ohm and of inductance is the henry.

It appears then if *CR* is measured in seconds

$$\text{farad} \times \text{ohm} = \text{second}$$

$$\text{Now farad} = \text{coulomb/volt}$$

$$\text{and ohm} = \text{volt/ampere}$$

so

$$\text{farad} \times \text{ohm} = \frac{\text{coulomb}}{\text{volt}} \times \frac{\text{volt}}{\text{ampere}}$$

$$= \frac{\text{coulomb}}{\text{ampere}}$$

and one ampere is one coulomb/second so

$$\text{farad} \times \text{ohm} = \frac{\text{coulomb}}{\text{coulomb}} \times \text{second}$$

and we see that farad × ohm does in fact give second. Similarly

$$\text{henry} = \frac{\text{volt} - \text{second}}{\text{ampere}}$$

$$\left(\text{from } e = -L\frac{di}{dt}, \text{ i.e. volt} = \frac{\text{henry} \times \text{ampere}}{\text{second}}\right),$$

so that

$$\text{henry/ohm} = \frac{\text{volt}}{\text{ampere}} \times \text{second} \times \frac{\text{ampere}}{\text{volt}}$$

$$= \text{second}$$

This approach to the checking of the correctness of a formula or equation is called *dimensional analysis*. It is often useful to return to basic principles and basic units to obtain a check in such cases.

Exponential changes: general equations

From the preceding theory we see that when a quantity is increasing exponentially the

$$\text{value at time } t = \text{steady state value} \times (1 - e^{-t/\tau})$$

where *t* is the time after the change begins and τ is the time constant.

When a quantity is decaying exponentially

$$\text{value at time } t = \text{steady state value} \times e^{-t/\tau}$$

It is sometimes easier to recall the general form of the equation and insert the particular relationships for the steady state value (*E* or *E/R*) and for the time constant (*CR* or *L/R*) as determined by the nature of the circuit.

Summary The current flowing in an inductor cannot change instantaneously owing to the opposing or back e.m.f. induced by the changing magnetic field in or around the inductor.

When an inductive circuit is closed the relationship between current I (amperes), the applied voltage E (volts) and the time t (seconds) which has elapsed after switching is given by

$$I = \frac{E}{R}(1 - e^{-Rt/L})$$

and between the inductor back e.m.f. V_L (volts) and the other quantities are

$$V_L = -E\,e^{-Rt/L}$$

where L is the circuit inductance (henrys) and R is the circuit resistance (ohms).

On opening an inductive circuit the collapsing inductor magnetic field again establishes a back e.m.f. which opposes changing current. On this occasion the current *decays* exponentially. The back e.m.f., after first rising instantaneously to a maximum value (the applied voltage if the circuit is in a steady state on switching), also decays exponentially. The equations are

$$I = \frac{E}{R}\,e^{-Rt/L}$$

and

$$V_L = E\,e^{-Rt/L}$$

The time constant in an inductive circuit, i.e. the time taken for any change to be complete if the initial rate of change were to remain the same and also the time after which the changing quantity has changed by 0.632 of its initial value, is equal to L/R seconds.

Example 4.7 A circuit is made up of a resistor and a second component connected in series. When the resistance of the resistor is increased the time taken after closing the circuit for the circuit to reach steady state is reduced. The second component:
   A. is the resistor;
   B. is an inductor;
   C. is a capacitor;
   D. cannot be identified without further information.

A. Circuits containing only resistance have no transient period. Steady state is reached immediately on closing the circuit so that the value of resistance has no effect.

B. The time constant of an *L-R* circuit is equal to *L-R*. If R is increased the time constant and thus the time required to reach steady state is reduced. This answer is therefore correct.

C. The time constant of a *C-R* circuit is equal to *CR*. If R is increased the time constant and thus the time required to reach steady state is increased. This answer is therefore incorrect.

D. This answer is incorrect; the nature of the second component can be identified by reasoning as in answer B above.

EXERCISE 4.1   1. A series *C-R* circuit has a capacitance of 1 μF and resistance of 450 Ω. Calculate the circuit time constant and the time taken from closing the circuit for the capacitor voltage to reach 47% of its final value.

2. A 50 V d.c. supply is applied to a circuit consisting of an 8 μF capacitor, a 12 μF capacitor, a 47 kΩ resistor and a 33 kΩ resistor connected in series. Calculate the time constant and the voltage across the 8 μF capacitor in the steady state.

3. The charging current in a circuit composed of a resistor and capacitor in series falls to 45 mA in 75 ms after applying 10 V d.c to the circuit. The circuit resistance is 120 Ω. Calculate the circuit capacitance.

4. Two capacitors of 12 μF and 8 μF are connected in parallel, the combination then being connected in series with a 1.5 kΩ resistor. Calculate the charging current in each capacitor 15 ms after applying 100 V d.c. to the circuit.

5. An inductive coil carries 0.6 A in the steady state when 150 V is applied across it. The current is 0.4 A after 45 ms has elapsed from the instant of closing the circuit. Calculate the coil inductance.

6. The current in a d.c. circuit composed of a 10 H coil and 100 Ω resistor in series reaches 40% of the maximum value 20 ms after closing the circuit. Calculate the coil resistance.

7. An inductive coil having a resistance of 150 Ω is connected to a 150 V d.c. supply. Steady state is reached 4 s after closing the circuit. Calculate the time after which the circuit current reaches 0.3 of the maximum value.

8. Calculate the back e.m.f. of a 10 H, 200 Ω coil 40 ms after a 100 V d.c. supply is connected to it.

Possible marks

SELF-ASSESSMENT EXERCISE 4   1. Define the time constant of a series *C-R* circuit.   (3)

2. Define the time constant of a series *L-R* circuit.   (3)

3. Sketch the capacitor voltage/time graph from zero to steady state when a d.c. supply is connected to a series *C-R* circuit.   (3)

4. A d.c. supply is connected to an inductive coil. Sketch the graph of back e.m.f. plotted against time showing the salient details.   (3)

5. Explain briefly why increasing the capacitance of a *C-R* circuit increases its time constant.   (3)

6. When 100 V is applied to a series *C-R* circuit the capacitor voltage reaches 63.2 V after 0.36 S. Calculate the circuit capacitance if the initial value of the charging current is 100 mA.   (5)

7. The time constant of a series *L-R* circuit is 0.02 s. The coil inductance is 5 H and the additional resistor has a resistance of 10 Ω. Calculate the coil resistance.   (5)

8. Three identical capacitors are connected in series with a 1000 Ω resistor. Steady state is reached 0.45 s after applying a direct voltage to the circuit. Calculate the capacitance of each capacitor.   (5)

9. Define time constant of an inductive circuit.

A 100 mH, 50 Ω coil is supplied with 20 V d.c. Calculate:
    (a) the time constant of the circuit;

(b) the time taken for the coil current to reach one half the final value;

(c) the final value of current. (14)

10. Explain why protective circuits or special switches are used for inductive circuits used with d.c. supplies.

A spark suppression circuit consists of placing a resistor across a 1 H, 150 Ω coil at the instant of opening the supply switch. Calculate the value of this resistor if the decay time of the current is to be reduced to 10% of the value of the rise time (to maximum current) of the circuit. (14)

11. Determine the value of resistor which when connected in series with (a) a 0.5 μF capacitor, (b) a 10 H 200 Ω inductor will allow the current in each of the two circuits to change by 50% of the maximum value within 20 ms from the instant of applying the supply voltage (14)

12. A direct voltage of 150 V is applied to a series *C-R* circuit of value 0.5 μF, 1 MΩ. Calculate:

(a) the capacitor voltage after 0.2 s, 0.5 s and 1.5 s;

(b) the initial charging current;

(c) the time constant;

(d) the value of capacitor which must be connected to the 0.5 μF capacitor in order to reduce the charging time to half its present value. How must this capacitor be connected? (14)

13. A square wave pulse of duration (a) 100 ms, (b) 1 ms is applied to a series *C-R* circuit of value 0.1 μF, 100 kΩ. Sketch the output voltage appearing across the capacitor in each case showing all relevant times and details of respective time constants. (14)

Answers

1. 0.45 ms; 0.286 ms

2. 0.384 s; 30 V

3. 1014.4 μF

4. 8 μF: 24.26 mA; 12 μF: 16.17 mA

5. 10.24 H

6. 255.4 Ω

7. 0.446 s

8. 44.93 V

SELF-ASSESSMENT EXERCISE 4

Marks

1. Definition: see text. (3)

2. Definition: see text. (3)

3. Sketch as fig. 4.1 (all salient points should be shown). (3)

4. Sketch as fig. 4.3 (note *back e.m.f.* graph is asked for). (3)

5. The greater the capacitance the greater the charge required per unit voltage applied. Time is required for the capacitor to charge and if the charging current remains the same, which it will do if the resistance remains the same, more time is required. Charging time is directly affected by time constant so that increasing the charging time by increasing the capacitance increases the time constant. (1) (1) (1)

6. Time taken to reach 63.2 V, i.e. 0.632×100 V (the maximum voltage)
$$= 0.36 \text{ s}$$
Therefore time constant = 0.36 S (2)

$$R = \frac{\text{voltage applied}}{\text{maximum charging current}} = 100 \text{ V}/100 \text{ mA} = 1 \text{ k}\Omega \qquad (1)$$

and $CR = 360 \mu F$ (1)

Therefore $C = 0.36/R = 0.36 \times 10^{-3}$
$$= 360 \mu F \qquad (1)$$

7. Time constant $= L/R \text{ s} = 0.02 \text{ s}$ (1)

$$= \frac{5}{R} = 0.02 \qquad (1)$$

so that $\qquad R = 5/0.02$
$$= 250 \ \Omega \text{ (total resistance)} \qquad (2)$$
Additional resistance $= 10 \ \Omega$
Coil resistance $= 250 - 10$
$$= 240 \ \Omega \qquad (1)$$

8. $5 CR = 0.45 \text{ s}$ (2)
Therefore $CR = 0.09 \text{ s}$ (1)

and $C = 0.09/1000$ (since $R = 1000 \ \Omega$)
$$= 90 \mu F \qquad (1)$$

There are three identical capacitors in series. Thus each capacitor is 3×90, i.e. $270 \mu F$. (1)

9. Definition: see text (3)

(a) Time constant $= L/R = 100/50 \text{ ms}$
$$= 2 \text{ ms} \qquad (3)$$

(b) $\qquad 0.5 \, I_{max} = I_{max} \, (1 - e^{-t/2 \times 10^{-3}})$ (2)

Therefore $\quad e^{-t/2 \times 10^{-3}} = 0.5$ (1)

$$e^{t/2 \times 10^{-3}} = 2 \qquad (1)$$

$$t = 2 \times 10^{-3} \ln 2$$
$$= 1.39 \text{ ms} \qquad (2)$$

(c) Final value of current $=$ applied voltage/resistance
$$= 20/50$$
$$= 0.4 \text{ A} \qquad (2)$$

10. On opening an inductive circuit the induced e.m.f. acts so as to maintain current flow in the circuit and through the opening switch. This causes sparking at and subsequent damage to the switch necessitating either protective circuits or special switches. (2) (2)

Rise time $= 5L/R$
$$= 5 \times 1/150 = 33.33 \text{ ms} \qquad (2)$$

Decay time with additional resistor $= 10\% \times 33.33$
$$= 3.33 \text{ ms} \qquad (2)$$

If new value of (total) resistance is R_1 then

$$5/R_1 = 3.33 \times 10^{-3} \, (L = 1\text{H}) \qquad (2)$$

$$\text{and } R_1 = \frac{5 \times 1}{3.33 \times 10^{-3}} = 1500 \ \Omega \qquad (2)$$

so that additional resistance
$$= 1500 - \text{coil resistance}$$
$$= 1350 \ \Omega \qquad (2)$$

11. (a) $0.5 = e^{-20 \times 10^{-3}/0.5 \times 10^{-6} \times R}$ (2)

where R is the series resistance so that

$$\frac{20 \times 10^{-3}}{0.5 \times 10^{-6} \times R} = \ln \frac{1}{0.5}$$ (2)

and hence $R = 57\,708\,\Omega$. (2)

(b) $0.5 = 1 - e^{-R \times 20 \times 10^{-3}/10}$ (2)

where R is the *total* circuit resistance so that

$$\frac{R \times 20 \times 10^{-3}}{10} = \ln \frac{1}{0.5}$$ (2)

$$\text{and } R = 346.6\,\Omega$$ (2)

The additional series resistance $= 346.6 - 200\,\Omega$
$$= 146.6\,\Omega$$ (2)

12. (a) After 0.2 S

$V_C = 150\,(1 - e^{-0.2/0.5 \times 10^{-6} \times 10^{6}})$
    $= 49.45$ V (2)

After 0.5 s

$V_C = 150\,(1 - e^{-0.5/0.5 \times 10^{-6} \times 10^{6}})$
    $= 94.82$ V (2)

After 1.5 s

$V_C = 150\,(1 - e^{-1.5/0.5 \times 10^{-6} \times 10^{6}})$
$= 142.53$ V (2)

(b) Initial charging current $= 150/10^6$ A
$\qquad\qquad\qquad\qquad = 0.15$ mA (2)

(c) Time constant $= 0.5 \times 10^{-6} \times 10^6$ s
$\qquad\qquad\quad = 0.5$ s (2)

(d) If the charging time $(5CR)$ is to be halved, R remaining the same then
the value of circuit capacitance must be halved to $0.25\ \mu\text{F}$. (2)
This may be achieved by adding an additional $0.5\ \mu\text{F}$ capacitor in *series* with
the $0.5\ \mu\text{F}$ capacitor. (2)

13. (a)

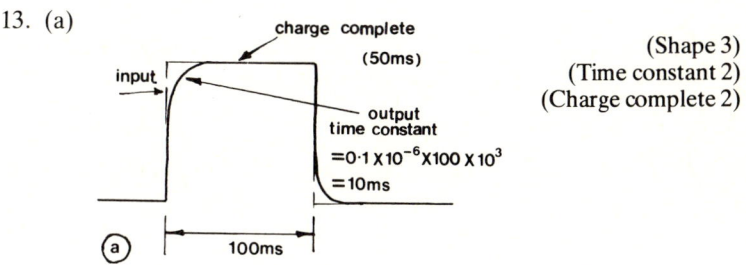

(Shape 3)
(Time constant 2)
(Charge complete 2)

(b)

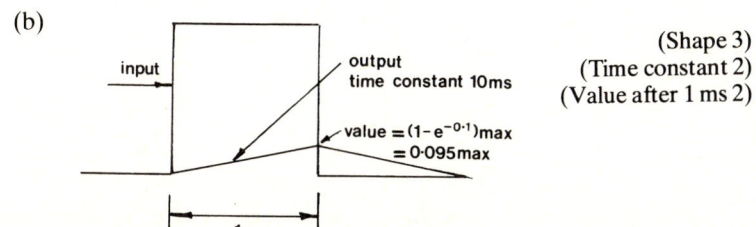

(Shape 3)
(Time constant 2)
(Value after 1 ms 2)

5 Transformers

Topic area: E

General objective

The expected learning outcome is that the student understands the principles of operation of transformers.

Specific objectives

The expected learning outcome is that the student:

6.1 *States the essential features of construction of power a.f. and r.f. transformers.*

6.2 *Draws the phasor diagram for an ideal transformer on no-load.*

6.3 *States that the volts-per-turn are constant and that the primary and secondary ampere-turns are equal in an ideal transformer.*

6.4 *States the equation $N_2/N_1 = V_2/V_1 = I_1/I_2$.*

6.5 *States that a transformer can be used to match a source to load, using the maximum power transfer theorem.*

6.6 *Outlines the derivation of the relation $R_2 = R_1 (N_2/N_1)^2$ for transformer matching.*

6.7 *States that iron losses consist of eddy-current and hysteresis losses in the core and differentiates between them.*

6.8 *Explains the choice of transformer core materials and construction to minimise core losses.*

6.9 *Calculates transformer per unit regulation, given secondary voltage on open circuit at rated load.*

Transformers – principles

A transformer consists of two or more coils linked by a magnetic circuit. An alternating supply applied to one coil, the primary winding, sets up an alternating current and thus a changing magnetic field. This field is linked to the other coil, the secondary winding, so that a voltage is induced across it. The induced voltage is then used as a supply. The main advantage is that the secondary voltage may be adjusted so that it is possible to obtain any level of voltage from a single supply. Transformers are widely used in distribution where the generated voltage is first increased to improve transmission efficiency and then progressively reduced to a safe working level for the consumer, in electronic equipment power supplies and in electronic amplifiers for matching.

Transformer construction

Transformers for use at power and audio frequencies (up to about 20 kHz) use metal cores for the magnetic circuit. At frequencies above this range dust cores and even air cores may be used because of the increasing losses at high frequencies.

Metal cores are made of high permeability steel alloys formed into thin laminations (about 0.35 mm thick) which are bolted together, each lamination being insulated from its neighbour. The purpose of a laminated rather than solid core is to reduce the effect of small circulating currents, called eddy currents, which are induced within the core by the changing flux. The core shape may be either of two shown in fig. 5.1. Part (a) is generally known as 'core construction', and part (b) is known as 'shell construction'. The majority of power transformer cores are of the form shown in fig. 5.1 a.

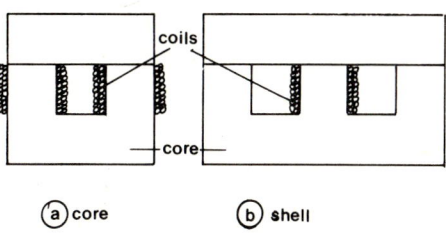

(a) core   (b) shell

Figure 5.1

Transformer windings are generally circular in cross-section to withstand the considerable mechanical stresses set up when on load. There are various types of winding depending upon the intended use of the transformer.

Once assembled, the windings and core are contained within a metal case or tank. The windings are cooled either by air or, on larger power transformers, by oil, and provision is made for the coolant to circulate through the container.

The ideal transformer

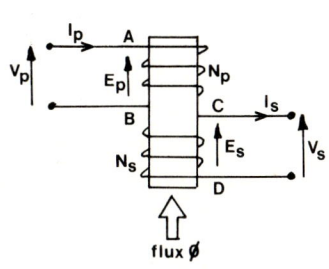

flux ϕ

Figure 5.2

If certain assumptions are made and the transformer is considered ideal, various basic relationships between voltages and currents at the primary and secondary sides of the transformer may be derived. In practice these relationships are approximate but nevertheless useful. For an ideal transformer the following assumptions are made:

(a) zero winding resistance;
(b) no flux losses, i.e. total flux links primary and secondary;
(c) zero core losses;
(d) the core permeability is extremely high so a negligible m.m.f. is required to set up the flux.

Consider the circuit of fig. 5.2 which shows an ideal transformer connected between a supply and a load. Provided that the flux has the same waveform as the current, which is assumed sinusoidal, the equation describing the flux may be written

$$\phi = \phi_{max} \sin 2\pi ft$$

where
ϕ is the instantaneous value of flux (Wb)
ϕ_{max} is the maximum value of flux (Wb)
f is frequency (Hz)
t is time (s)

Kirchhoff's voltage law applied to the primary circuit indicates that the voltage E_p induced in the primary winding by the changing flux, is equal to and opposes the applied voltage V_p.

From Faraday's Law this induced voltage is equal to the number of turns times the rate of change of flux. Differentiating the equation gives

$$\text{rate of change of flux with time} = 2\pi f\phi_{max}\cos 2\pi ft$$

so that

$$E_p = N_p \times 2\pi f\,\phi_{max}\cos 2\pi ft$$

which is a cosinusoidal wave, i.e. a sinusoidal wave displaced from the flux wave by $\pi/2$ rad leading, with an amplitude $2\pi fN_p\,\phi_{max}$.

The r.m.s. value of this voltage $= \dfrac{1}{\sqrt{2}}$ peak value

thus

$$E_p = \frac{1}{\sqrt{2}}2\pi fN_p\,\phi_{max}$$

$$= 4.44fN_p\,\phi_{max}$$

and since

$$E_p = V_p$$

$$V_p = 4.44fN\phi_{max}$$

where E_p and V_p are the r.m.s. values of primary induced voltage and applied voltage, respectively. It follows that

$$\phi_{max} = V_p/4.44fN_p$$

The changing flux links the secondary winding and the secondary voltage E_s is induced. From the above it follows that

$$E_s = 4.44fN_s\,\phi_{max}$$

Neglecting winding impedances so that

$$E_s = V_s$$
$$V_s = 4.44fN_s\,\phi_{max}$$

and substituting for ϕ_{max} using the relationship derived earlier

$$V_s = 4.44fN_s\,\frac{V_p}{4.44fN_p}$$

$$= \frac{N_s}{N_p}V_p$$

thus

$$\frac{V_s}{V_p} = \frac{N_s}{N_p}$$

The ratio of secondary to primary voltage is therefore the ratio of the secondary turns to primary turns. In practice, when these assumptions are not made this is a very close approximation. One of the main advantages of the transformer is that any value of secondary voltage may be obtained by adjusting the turns ratio.

Lenz's Law states that the voltage induced in the secondary acts in a direction so as to oppose what is causing it. This means that when current is drawn from the secondary winding the m.m.f. set up by the secondary current will oppose the main flux causing the induction. Kirchhoff's Law applied to the primary circuit states that the primary induced voltage, which is also determined by the main flux, is equal and opposite to the applied voltage. If the main flux were reduced by the demagnetising effect of the secondary current this would reduce the primary induced voltage and Kirchhoff's Law would not hold. What in fact happens is that when secondary current is drawn a primary current flows of such a value that its m.m.f. opposes the demagnetising m.m.f. of the secondary so that the main flux remains unaffected and Kirchhoff's Law is still valid. It should be remembered at this point that the m.m.f. causing the main flux is assumed negligible because of the extremely high permeability of the core. In practice a small component of the primary current is in fact required to establish the main flux.

The demagnetising m.m.f. of the secondary winding is equal to the current-turns product, i.e. $I_s N_s$ using the symbols above. The compensating primary m.m.f. (neglecting the m.m.f. to set up the main flux) is $I_p N_p$. Thus

$$I_p N_p = I_s N_s$$

and

$$\frac{I_s}{I_p} = \frac{N_p}{N_s}$$

which is again a close approximation when the 'ideal' assumptions are not made. Note that the secondary: primary current ratio is the primary:secondary turns ratio, i.e. the inverse of the voltage relationship.

It is sometimes wrongly implied that the opposing nature of the induced voltage across the secondary leads to it being in antiphase with the applied primary voltage. This is not so since the phase relationship between primary and secondary voltages of a transformer is determined by which ends of the windings are taken as reference. For example, in the circuit of fig. 5.2 the potential at point C with reference to point D is rising and falling in phase with the potential at point A with reference to B. Alternatively, the potential at point D with reference to point C is changing in antiphase with the potential at point A with reference to point B. It is thus possible to obtain a zero phase shift or a 180° phase shift from primary to secondary by the appropriate connection of the windings relative to one another. The opposing nature of the induced voltages thus shows itself only in the establishment of the m.m.f. as described above and not in the phase shift, if any, between windings.

Example 5.1 The primary voltage and secondary current of a transformer having a 50:1 turns ratio are 200 V and 1.5 A, respectively. Calculate the primary current and secondary voltage.

Using the equations already derived

$$\frac{V_p}{V_s} = \frac{N_p}{N_s} = \frac{I_s}{I_p}$$

so that

$$I_p = \frac{N_s}{N_p} I_s$$

$$= \frac{1}{50} \times 1.5$$

$$= 30\,\text{mA}$$

and

$$V_s = \frac{N_s}{N_p} V_p$$

$$= \frac{1}{50} \times 200$$

$$= 4\,\text{V}$$

The primary current and secondary voltage are 30 mA and 4 V, respectively.

Example 5.2   The primary voltage and current of a transformer having a 10:1 turns ratio are 200 V, 1 A, respectively. The secondary voltage and current are respectively
  A. 2000 V; 10 A
  B. 20 V; 100 mA
  C. 20 V; 10 A
  D. 2000 V; 100 mA

The secondary voltage

$$V_s = \frac{N_s}{N_p} V_p$$

$$= \frac{200}{10}$$

$$= 20\,\text{V}$$

The secondary current

$$I_s = \frac{N_p}{N_s} I_p$$

$$= 10 \times 1$$

$$= 10\,\text{A}$$

The correct answer is thus 20 V, 10 A and option C is the correct choice.

In option A both primary voltage and current have been *multiplied* by the turns ratio; the current is correct but the voltage is not.

In option B both primary voltage and current have been *divided* by the turns ratio, which gives the correct voltage but the incorrect current.

In option D the errors of A and B have been compounded, the primary voltage being multiplied by the turns ratio, the primary current being divided by the turns ratio. This is of course the exact opposite to the correct method and results in two wrong answers.

Impedance matching

The equations of the ideal transformer may be used to show another useful characteristic, that of impedance matching.

It has been shown that

$$V_p = \frac{N_p}{N_s} V_s$$

and

$$I_p = \frac{N_s}{N_p} I_s$$

Dividing

$$\frac{V_p}{I_p} = \left(\frac{N_p}{N_s}\right)^2 \frac{V_s}{I_s}$$

but

$$\frac{V_s}{I_s} = Z_s$$

(where Z_s is the secondary impedance), so that

$$\frac{V_p}{I_p} = \left(\frac{N_p}{N_s}\right)^2 Z_s$$

In the circuit of fig. 5.2 the ratio V_p/I_p represents the impedance into which the source generator is working. It is this effective impedance which determines the current supplied by the source, i.e. the effective impedance is the load as 'seen' by the source so that,

$$\text{effective load} = \left(\frac{N_p}{N_s}\right)^2 Z_s$$

The circuit of fig. 5.2 assumed a source having zero internal impedance. In practice where the internal impedance is finite the current supplied by the source will be determined by the internal impedance and effective impedance in series (see fig. 5.3).

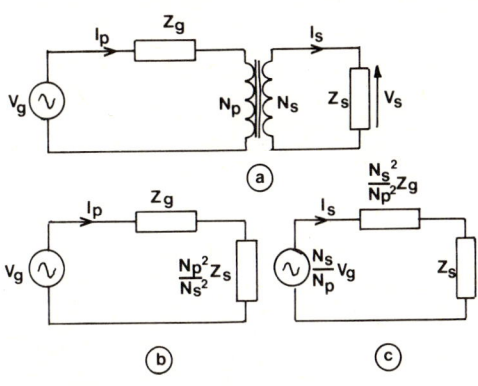

Figure 5.3

Determination of the effective impedance using the above equation is known as transferring the load impedance from secondary to primary. The maximum power transfer theorem (Chapter 1) states that, for transfer of maximum power from a source to load, the source resistance and load resistance should be equal. By insertion of a transformer between a source and load the effective load resistance may be made equal to the source resistance by a suitable choice of transformer turns ratio. This method of 'matching source to load' is commonly used in power amplifier circuits.

Note that as impedances, voltages and currents can be transferred from secondary to primary, the reverse process also applies. If all quantities in fig. 5.3(b) are transferred to secondary, the generator appears as $(N_s/N_p) \times V_g$ the source impedance as $(N_s/N_p)^2 \times Z_g$ and the load impedance would remain at Z_s.

The current I_p in the circuit of fig. 5.32(b) is given by

$$I_p = V_g/[Z_g + (N_p/N_s)^2 Z_s]$$
$$= N_s^2 V_g/(N_s^2 Z_g + N_p^2 Z_s)$$

The current I_s in the circuit of fig. 5.3(c) is given by

$$I_s = \frac{(N_s/N_p)V_g}{(N_s/N_p)^2 Z_g \times Z_s}$$

$$= \frac{N_p}{N_s} \left(\frac{N_s^2 V_g}{N_s^2 Z_g + N_p^2 Z_s} \right)$$

$$= \frac{N_p}{N_s} I_p$$

which is the result obtained earlier.

Example 5.3 A voltage source having a sinusoidal voltage waveform of amplitude 2 V and an internal resistance of 40 Ω is connected via a transformer to a 4000 Ω load. Calculate:

(a) the turns ratio of the transformer for maximum power transfer between source and load;

(b) the load current, voltage and power under these conditions.

(a) For maximum power transfer the effective (transferred) load resistance should equal the source resistance of 40 Ω. From the above theory

$$40 = \left(\frac{N_p}{N_s} \right)^2 4000$$

and

$$\frac{N_p}{N_s} = \sqrt{\frac{40}{4000}}$$

$$= \frac{1}{10}$$

The turns ratio, primary:secondary, is thus 1:10.

(b) The secondary voltage, V_s, is given by

$$V_s = \frac{N_s}{N_p} V_p$$

$$= 10 \times 2$$

$$= 20 \text{ V (peak)}$$

The secondary current may be obtained from the equation $I_p = N_s/N_p$. I_s after first determining the primary current using an equivalent circuit of the form shown in fig. 5.3b. Alternatively, it may be obtained directly from a circuit of the form shown in fig. 5.4a.

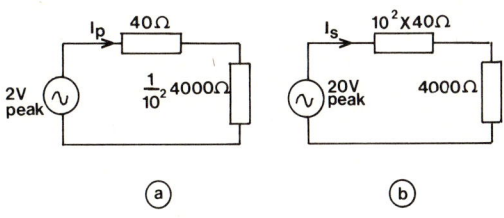

Figure 5.4

From this circuit

$$I_p = 2/80$$
$$= 0.025 \text{ A}$$

so that

$$I_s = 0.0025 \text{ A}$$

Alternatively, from fig. 5.4b

$$I_s = 20/8000$$
$$= 0.0025 \text{ A}$$

as before.

$$\text{The power in the load} = \left(\frac{0.0025}{\sqrt{2}}\right)^2 \times 4000$$

(note that the r.m.s. value must be used)
$$= 12.5 \text{ mW}$$

This is the maximum power that it is possible to transfer from this particular source.

The transformer off load

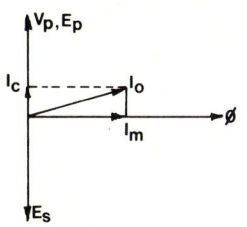

Figure 5.5

The ideal transformer assumes that the m.m.f. required to set up the main flux is negligible and the core losses are zero. This implies that on no load no primary current flows since a primary current is only required to set up the compensating m.m.f. for the demagnetising effect of the m.m.f. produced by the secondary (load) current.

In practice an m.m.f. is required for the main flux and the core losses are finite. On no load, then, a small primary current flows, designated I_0 in the phasor diagram of fig. 5.5.

The no load current is displaced from the applied voltage by almost $\pi/2$ rads (compare the phasor diagram of a pure inductor) and is almost in phase with the flux phasor. The no load current may be considered to have two components, I_c, in phase with the applied voltage, the product $V_p I_c$ being the power loss within the magnetic core due to hysteresis and eddy currents (see below), and a component I_m in phase with and producing the main flux ϕ. The primary induced voltage E_p is in phase with V_p and leads the flux causing it by $\pi/2$ rad. This diagram is somewhat simplified in that no account is taken of the primary winding impedance (in more detail V_p is actually composed of two components, one equal and opposite to E_p and one being the voltage drop due to finite winding impedance).

In the phasor diagram of fig. 5.5, E_s is shown in antiphase with V_p but it can equally be in phase as was described above. For simplification the turns ratio has been taken as unity so that E_p and E_s are shown equal. The phasor diagram for the ideal transformer on no load is fig. 5.5 with I_c, I_0 and I_m omitted.

Example 5.4   An ideal transformer having a primary winding of 2000 turns has a 25 Ω resistive load across the secondary winding. An alternating voltage of 100 V at 50 Hz input produces 75 V across the load. Calculate:

  (a) the number of turns of the secondary winding;
  (b) the equivalent input resistance of the transformer;
  (c) the current in each winding when on load.

(a) The ratio of primary: secondary voltage

$$= 100/75$$

Thus the ratio of primary: secondary turns

$$= 100/75$$

Primary turns

$$= 2000$$

Therefore

$$\text{secondary turns} = \frac{75}{100} \times 2000$$

$$= 1500$$

(b) Equivalent input resistance of the transformer is the resistance 'seen' by a source connected to it, i.e. the load resistance referred to the primary;

$$\text{input resistance} = \left(\frac{N_p}{N_s}\right)^2 Z_s$$

$$= (100/75)^2 \times 25$$
$$= 44.4 \ \Omega$$

(c) Secondary current $= \dfrac{\text{secondary voltage}}{\text{load}}$

$$= 75/25$$
$$= 3 \text{ A}$$

$$\text{primary current} = \frac{N_s}{N_p} \times I_s$$

$$= \frac{75}{100} \times 3$$

$$= 2.25 \text{ A}$$

Per unit regulation

As the secondary load on a transformer is increased (i.e. the secondary current rises), the output voltage falls from the no load value. The voltage regulation of a transformer is the change which occurs in the output voltage as the load is increased from no load to full load at the same power factor. This change when divided by the no load voltage is called the *per unit regulation*. The per unit method is often used with other quantities related to transformers or electrical machines, the divisor or base being a preselected value, in this case the no-load secondary voltage.

Example 5.5   The output voltage of a transformer on no load is 1000 V. At full load the output voltage is 950 V. Calculate the per unit regulation.

$$\text{Change in output voltage} = 1000 - 950$$
$$= 50 \text{ V}$$

The p.u. regulation is then 50/1000, i.e. 0.05.

Transformer efficiency and losses

The efficiency of modern transformers is very high indeed, being of the order of 95–99%. However, the losses can never be made entirely negligible. Transformer losses may be subdivided into two types: iron losses and copper losses.

Iron losses

These losses may be further divided into hysteresis losses and eddy current losses. The hysteresis loss is common to any magnetic material subjected to a cycling flux. When the magnetising force is reduced to zero the flux density remains finite and the magnetising force must be reversed and increased in the opposite direction to reduce the flux density to zero. The hysteresis (lagging behind) of the flux density when the magnetising force is changed is due to the shifting position of so called 'magnetic domains' within the material. These may be visualised as small bar magnets within the magnet. If the m.m.f. varies cyclically, the continuous movement of the domains results in a loss in energy which displays itself as heat. Hysteresis loss varies with the material type as shown in fig. 5.6 which shows several typical plots of flux density/magnetising force curves for different materials. It can be shown that the energy loss is proportional to the area contained within these curves. The curves are known as *B/H* loops. Hysteresis loss is proportional to the frequency of alternation.

Iron losses in general may be reduced by the choice of a low loss core material, i.e. having a small-area *B/H* loop, to reduce

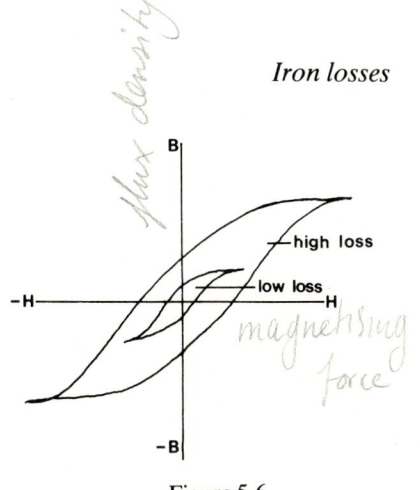

Figure 5.6

hysteresis losses, and by laminating the core, as described earlier, to reduce and restrict the eddy current paths within the core.

The main flux within a transformer varies only slightly between no load and full load and so iron losses may normally be assumed constant. They may be measured by determination of the input power to the transformer (using a wattmeter) when the secondary circuit is open circuited. The copper losses, as described below, may be considered negligible under these conditions.

Copper losses   These are the normal resistance power losses caused when current flows through the windings. An alternative name for them is the I^2R loss, the name being self-explanatory. Copper losses at full load current may be measured by short circuiting the secondary and applying a low voltage to the primary of sufficient value to circulate full load secondary current. Under these conditions the iron loss is fairly small compared to the copper loss and a wattmeter connected at the input may be assumed to measure the full load copper losses only.

The efficiency of a transformer is given by

$$= \frac{\text{output power}}{\text{input power}} \times 100\%$$

$$\frac{\text{input power} - \text{losses}}{\text{input power}} \times 100\%$$

$$1 - \frac{\text{losses}}{\text{input power}} \times 100\%$$

$$1 - \frac{\text{iron loss} + \text{primary copper loss} + \text{secondary copper loss}}{\text{input power}} \times 100\%$$

It can be shown that maximum efficiency of a transformer occurs when the (variable) copper losses are equal to the (constant) iron losses.

Example 5.6   A 150 kVA transformer has an iron loss of 700 W and a full load copper loss of 1800 W. Calculate:
(a) the efficiency at full load at 0.8 power factor;
(b) the maximum efficiency and the kVA output at which this occurs, assuming the same power factor.

(a)   input power = input volt-amperes × power factor
$$= 150 \times 0.8 \, \text{kW}$$
$$= 120 \, \text{kW}$$
iron losses = 700 W
copper losses = 1800 W

Hence
$$\text{efficiency} = \left(1 - \frac{700 + 1800}{120\,000}\right) 100\%$$
$$= \overline{1 - 0.0208} \; 100\%$$
$$= 97.92\%$$

(b) Maximum efficiency occurs when

$$\text{copper losses} = \text{iron losses}$$

In this case when

$$\text{copper losses} = 700 \text{ W}$$

or

$$\text{total loss} = 1400 \text{ W}$$

The efficiency under these conditions

$$= \left(1 - \frac{1400}{120\,000}\right) 100\%$$

$$= 98.83\%$$

Output kVA

$$= 0.9883 \times 120$$
$$= 118.6 \text{ kVA}$$

Example 5.7 A single-phase transformer is rated at 10 kVA 230/ 100 V. When the secondary terminals are open circuited and the primary winding is supplied at normal voltage (230 V), the current input is 2.6 A at a power factor 0.3. When the secondary terminals are short-circuited and a reduced primary voltage causes the full load current to flow in the secondary, the primary power input is 240 W. Calculate:

(a) the efficiency at full load, unity power factor;
(b) the value of the maximum efficiency.

(a) From the open circuit test,

$$\text{iron losses} = 230 \times 2.6 \times 0.3$$
$$= 179.4 \text{ W}$$

From the short circuit test,

$$\text{copper losses} = 240 \text{ W}$$
$$\text{total losses} = 419.4 \text{ W}$$

$$\text{efficiency} = \left(1 - \frac{419.4}{10\,000}\right) 100\%$$

(The input is 10 000 W since the power factor is unity)

$$= 95.8\%$$

(b) Maximum efficiency occurs when

$$\text{copper loss} = \text{iron loss}$$

i.e.

$$\text{total losses} = 2 \times \text{iron loss}$$
$$= 358.8 \text{ W}$$

$$\text{Maximum efficiency} = \left(1 - \frac{358.8}{10\,000}\right) 100\%$$

$$= 96.41\%$$

Summary

A transformer consists of two or more coils linked by a magnetic circuit. An input voltage is applied to one coil, called the primary winding, and one or more voltages are taken from the other coil(s), called the secondary winding(s).

The general equations are

$$\frac{V_s}{V_p} = \frac{N_s}{N_p} = \frac{I_p}{I_s}$$

where V_p is the primary voltage, V_s is the secondary voltage, I_p the primary current, I_s the secondary current and N_p, N_s the primary and secondary windings number of turns, respectively.

A transformer may be used to match impedances, the relevant equation being

$$\text{effective load} = \left(\frac{N_p}{N_s}\right)^2 Z_s$$

where 'effective load' is that 'seen' by the source and Z_s is the actual secondary load impedance.

Transformer losses are copper losses, due to the resistance of the windings, and iron losses due to eddy current loss and magnet circuit or hysteresis loss. Transformer efficiency is given by

$$\left[1 - \left(\frac{\text{iron loss} + \text{copper losses}}{\text{input power}}\right)\right] \times 100\%$$

Maximum efficiency occurs when iron losses and copper losses are equal.

SELF-ASSESSMENT EXERCISE 5

Possible marks

1. State the equation relating turns ratio of a transformer with its primary and secondary voltages and currents. (3)

2. State the relationship between effective load as seen by a source connected to a transformer and the actual load on the secondary side. (3)

3. State the equation for efficiency of a transformer in terms of its losses. (3)

4. Define transformer losses. (3)

5. When does maximum efficiency of a transformer occur? (3)

6. A transformer of turns ratio 100:1 has an input of 100 V at 0.05 A. Calculate the output voltage and current. (5)

7. A transformer of turns ratio 5:1 is connected on its secondary side to a 150 Ω load. What is the effective load seen by the primary source? (5)

8. Calculate the maximum efficiency of a 100 kVA transformer having iron losses of 575 W. (5)

9. Describe typical forms of transformer construction for use at various ranges of frequency. (14)

10. What are the main assumptions made in the ideal transformer?
A practical transformer has an input voltage and current of 240 V, 1 A respectively, an efficiency of 98% and an output voltage of 20 V.
Calculate:
(a) the turns ratio;
(b) the input and output power assuming unity power factor;
(c) the secondary current. (14)

11. Describe a method of impedance matching using a transformer. Derive any equation included in your description. It is required to use a transformer to match into a 400 Ω resistive load the source resistance being 4 Ω. Calculate the transformer turns ratio. (14)

12. List the main sources of loss in a transformer and describe methods of reducing losses.
A 500 kVA transformer has an iron loss of 1.2 kW and a full load copper loss of 2.5 kW. Calculate the efficiency at full load (a) assuming unity power factor (b) with a 0.8 power factor. Determine the maximum efficiency of the transformer at unity power factor. (14)

13. Describe methods of determining transformer losses.
A 150 kVA transformer has iron losses of 1.4 kW and copper losses at full load of 2.8 kW. Calculate the efficiency of the transformer at full load, the maximum efficiency of the transformer and the output power at the maximum level of efficiency. Assume unity power factor. (14)

Answers

Marks

SELF-ASSESSMENT EXERCISE 5 1.

$$\frac{V_p}{V_s} = \frac{N_p}{N_s} = \frac{I_s}{I_p}$$ (1½ each)

2. effective load $= (N_p/N_s)^2 \, Z_s$ (3)

3.

$$\left(1 - \frac{\text{iron losses} + \text{copper losses}}{\text{input power}}\right) \times 100\%$$ (3)

4. Definition of copper loss, iron losses (both kinds to be described). (1 each)

5. When copper losses and iron losses are equal. (3)

6. 1 V, 5 A (2½ each)

7. Effective load $= 5^2 \times 150$ (3)
$= 3.75 \, k\Omega$ (2)

8. Maximum efficiency

$$= \left(1 - \frac{2 \times 575}{100\,000}\right) 100\%$$ (3)

$$= 98.85\%$$ (2)

9. See text for full description. Answer should include type of core (as determined by frequency), laminations, eddy currents, construction (core and shell) and coolant. (core 2)
(laminations 2)
(eddy currents 2)
(construction 6)
(coolant 2)

10. See text (four assumptions).   (1½ each)

(a) Turns ratio = 240/20   (20)
          = 12:1   (1)

(b) Input power = 240 × 1   (1)
            = 240 W   (1)
Output power = 0.98 × 240   (1)
            = 235.2 W   (1)

(c) Secondary current = 12 × 1
              = 12 A   (1)

11. Description of use of transformer between source and load to present an effective load equal to the source resistance. Derivation of formula.
                                                                   (5) (4)

$$4 = \left(\frac{N_p}{N_s}\right)^2 400$$ (3)

$$\frac{N_p}{N_s} = \sqrt{\frac{4}{400}} = \frac{1}{10}$$ (2)

12. See text. List to include methods of reducing.
   copper   (1)
   iron (eddy current)   (1)
   iron (core)   (1)
   laminations   (1)
   low loss core   (1)
   low resistance windings   (1)

Efficiency at unity p.f.

$$= \left(1 - \frac{3.7 \times 1000}{500\,000}\right) \times 100\%$$ (1)

$$= 99.26\%$$ (1)

Efficiency at 0.8 p.f.

$$= \left(1 - \frac{3.7 \times 1000}{500\,000 \times 0.8}\right) \times 100\%$$ (1)

$$= 99.1\%$$ (1)

Maximum efficiency (when copper loss = 1.2 kW)   (2)

$$= \left(1 - \frac{2 \times 1.2 \times 1000}{500\,000}\right) \times 1000$$ (1)

$$= 99.52\%$$ (1)

13. See text.   (open circuit test 3)
                                                          (short circuit test 3)

$$\text{Efficiency} = \left(1 - \frac{4.2}{150}\right) \times 100\%$$ (2)

$$= 97.2\%$$ (1)

$$\text{Maximum efficiency} = \left(1 - \frac{2.8}{150}\right) \times 100\%$$ (2)

$$= 98.13$$ (1)

Output power = 0.9813 × 150 × 10^3 W   (1)
          = 147.2 kW   (1)

6 Machines

Topic area: F

General objective   *The expected learning outcome is that the student understands the principles of rotating machines.*

Specific objectives   *The expected learning outcome is that the student:*
7.1   *Explains that:*
    (a) Motors convert electrical energy into mechanical energy.
    (b) Generators convert mechanical energy into electrical energy.
7.2   *Labels on a given diagram the essential parts of a d.c. machine.*
7.3   *Explains how*
$$E = Blv \text{ and } F = BlI$$
    apply to both motors and generators.
7.4   *Describes with the aid of diagrams the rectifying inverting action of a simple commutator.*
7.5   *Explains that:*
    (a) $E = K_e \phi N$
    (b) $T = K_t I_a \phi$
    (c) $E = V \pm$ (internal volt drop)
    apply to motors and generators.
7.13   *Solves problems using the equation of 7.5.*

Principles of d.c. machines   Electrical machines may be categorised in terms of the type of supply applied to or taken from them, i.e. alternating current or direct current and also by the main function of the machine, i.e. whether it is a motor or generator. There are then four main types: d.c. motor, d.c. generator, a.c. motor and a.c. generator (or alternator).

Certain basic principles, however, apply to all machines. They are:

(a) all rotating machines convert energy from one form to another: motors convert electrical energy to mechanical energy (of movement), generators convert mechanical energy to electrical energy;

(b) whenever a magnetic field surrounding, or sufficiently near to, a conductor changes, an e.m.f. is induced across the conductor (Faraday's Law of electromagnetic induction);

(c) magnetic fields exert a force on each other, the force may be one of attraction or one of repulsion as determined by the relative directions of action of the magnetic fields concerned.

Basic principle (a) will be examined later when we consider the relationship between the input and output energies and machine efficiency. Basic principles (b) and (c) may be written mathematically as follows.

The e.m.f. E (volts) induced across a conductor of length l (metres) moving through a magnetic field of flux density B (tesla) at a velocity of v (metres/second) is given by

$$E = Blv$$

and when a conductor of length l (metres) carrying a current I (amperes) is situated in a magnetic field of flux density B (tesla) the force experienced by the conductor (due to interaction of the main magnetic field and the magnetic field set up by the current) F (newtons) is given by

$$F = BlI$$

The first equation comes from a consideration of Faraday's Law of electromagnetic induction, which states for a single turn of conductor situated in a magnetic field changing at the rate of $d\phi/dt$ webers/second, the induced e.m.f. E (volts) is given by

$$E = \frac{d\phi}{dt}$$

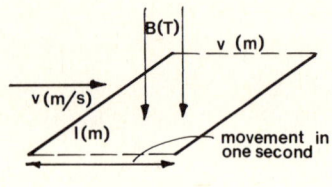

Figure 6.1

If we consider a conductor of length l metres moving at v metres/second then in one second it will have moved through an area of magnetic flux equal to lv square metres (see fig. 6.1). If the flux density is B webers/square metre (tesla) then in one second the flux swept through by the conductor will be $B \times$ area, i.e. Blv webers. The rate of change of flux is therefore Blv webers/second and the induced e.m.f.

$$E = Blv$$

as stated earlier.

The second mathematical relationship was considered in *Electrical and Electronic Principles 2*. The force F is directly proportional to each of the variables length l, current I and flux density B and if SI units are used the constant which is needed to convert the statement of proportionality

$$F \propto BlI$$

into the equation

$$F = BlI$$

is unity.

The easiest way to determine the direction of action of an induced voltage or of the force set up between magnetic fields is to use the simple principles

like fields repel
unlike fields attract

and whenever a voltage is induced it acts in a direction *so as to oppose what is causing it* (Lenz's Law). See pages 60–65 in *Electrical and Electronic Principles 2*.

D.C. machines' construction

The simplest form of construction of an electrical machine consists of a single coil which is free to rotate within the magnetic field set up by a permanent magnet as shown in fig. 6.2. If this simple machine is

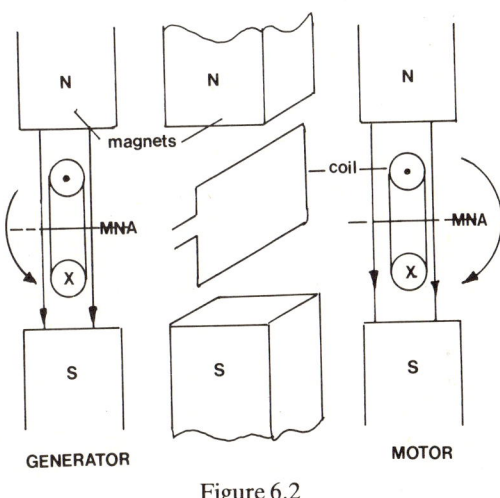

GENERATOR   MOTOR

Figure 6.2

to be used as a d.c. generator or motor it has a number of disadvantages:

(a) The magnet circuit has an extremely high reluctance due to the airgap which is itself not of uniform size (it is smaller in the centre of each pole piece than at its edges).

(b) If the machine is to be used as a generator large power outputs cannot be obtained from a single coil; similarly if it is to be used as a motor large power inputs cannot be made to a single coil and the machine would provide only a very small torque (turning force).

(c) The magnetic field is from permanent magnets which are prone to age, subsequently reducing the field strength. The magnetic field is also not directly controllable.

(d) The e.m.f. induced in a single conductor both changes direction and varies considerably in size as the coil rotates, being highest when the magnetic flux cut per second is highest. This occurs as the conductor is moving past each pole piece. When the conductor is midway between poles (at position MNA) it is moving *parallel* to the flux and is not in fact cutting it all. Here the induced e.m.f. is zero. The direction of action of the induced voltage is in one direction when it is passing one pole piece and in the opposite direction when it is passing the other pole piece.

In a practical machine the first disadvantage is overcome by use of as small an airgap as possible together with specially shaped pole pieces, as shown in fig. 6.3, to maintain an airgap of uniform width.

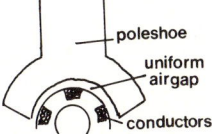

Figure 6.3

The second disadvantage of a single-coil arrangement is to use a large number of conductors suitably connected in series and/or parallel so that higher voltages and currents may be taken from or applied to the machine, depending upon whether it is used as a generator or as a motor. The conductors when connected are known collectively as the *armature*.

A non-controllable magnetic field which varies as the machine ages is avoided by using *field coils* to establish the main machine field. With this arrangement the strength of the main field may be altered at will by adjustment of the field-coil current.

The production of the main magnetic field is called *excitation* of the machine and machines may have their field coils connected so that they are separately excited or self-excited, see fig. 6.4. In separately excited machines the field coils are not connected in any way to the armature circuit; in self-excited machines the field coils may be connected in series, in parallel or partly in series and partly in parallel with the armature windings. The term 'self-excited' applies strictly only to generators (since they produce an e.m.f. which is used for the field coils as well as the load); for the motor the term means that the same source of (external) supply is used for the field coils as well as for the armature.

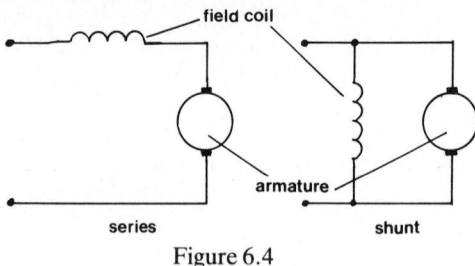

Figure 6.4

Fig. 6.4 shows different methods of connection, the parallel connection being also called a shunt connection. Part-series/part-parallel connection is 'compounding' and here the different parts of the field coils may be connected so as to assist each other (cumulative compounding) or oppose each other (differential compounding). The various methods of connection affect the characteristics of the machine which are discussed later.

The variation in magnitude and direction of the voltage induced in each conductor as the coil rotates in the magnetic field cannot be altered as such, but by using a large number of conductors suitably connected to each other and to the external load a direct voltage of virtually constant magnitude may be obtained. The external connections are made via a split-connector called a commutator, a simple form of which is shown in fig. 6.5. Here the top brush is always connected to the conductor moving past the top pole, and the bottom brush is always connected to the conductor moving past the bottom pole. Thus in the arrangement shown, with the machine being used as a generator and the direction of rotation being anticlockwise, when a load is connected to the commutator, current flows *from* the top brush through the load and *to* the bottom brush,

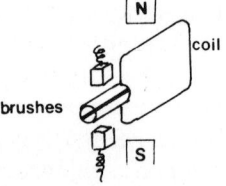

Figure 6.5

i.e. the top brush is the positive side of the supply produced by this machine, as far as it affects an external load, the bottom brush is the negative side of the supply.

If the machine is to be used as a motor the current flowing *into* a conductor from the external supply must always be in the same direction for a particular position of the conductor, for example, to produce an anticlockwise rotation the current in the conductor as it moves past the top pole must always flow into the conductor and as it moves past the bottom pole must always flow out of the conductor. If each conductor is always connected to a particular side of the supply this would not occur and the force produced by the interaction between the main field and the conductor field would act in one direction as the conductor moves past one pole and in the opposite direction as the conductor moves past the other pole. The commutator arrangement ensures that as each conductor moves into a new position it is connected to the appropriate side of the supply to maintain the same direction of action of the force on the conductor in that position and to ensure the same direction of action of the overall torque. The commutator is useful then, whether the machine is to be used as a d.c. generator or motor.

As it stands the output from a single coil two segment commutator in a generator is now unidirectional but is still of course fluctuating in size as the conductor moves from the zero e.m.f. position (between poles) to the maximum e.m.f. position (directly under a pole) to the zero position again. This fluctuation may be reduced to virtually zero by using many conductors and a multi-segment commutator as shown in fig. 6.6. Here the maximum induced e.m.f. is 'picked off' from each conductor as it moves under the pole pieces.

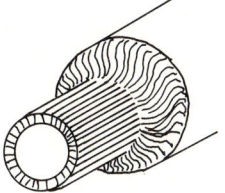

Figure 6.6

For good commutation the brushes should be connected at any instant to conductors in which no e.m.f. is being induced (at the instant of passing the brush) and for this reason the brushes should be sited along the magnetic neutral axis, shown as MNA in fig. 6.2 (here conductors are moving parallel to the field and not cutting it). The MNA in fact moves when the machine is loaded due to a phenomenon called *armature reaction* and the brush position should be moved accordingly. This is discussed later. If the point concerning no e.m.f. in the conductor at the brush is confusing it should be noted that although there may be no e.m.f. induced in the particular conductor in contact with the brush, the conductor is internally connected to other conductors in which an e.m.f. *is* being induced so the brushes do still operate.

In a practical machine the part which rotates is called the *rotor*. The arrangement of conductors on the rotor is called the armature of the machine. That part of the machine which is stationary is called the machine *stator*.

General equations

A generator is supplied with two inputs, a magnetic field and a turning force, or torque, on the rotor, the generator output being voltage. A motor is supplied with a magnetic field and an applied voltage, the output being the turning force, or torque, of the rotor. In both kinds of machine there are distinct similarities, however, and certain equations may be derived which apply to either kind of machine.

Induced and terminal voltage and
voltage drop

In both motors and generators the armature conductors are subjected to a changing magnetic field. In the generator the changing magnetic field is brought about by driving the generator, causing the armature conductors to move through the field (or in large a.c. machines causing the magnetic field to move past stationary conductors). In the motor, once the rotor is moving, a similar situation applies, on this occasion the motion causing the changing field being due to the motor itself. In both kinds of machine, therefore, there is an induced e.m.f. across the armature. In the generator this induced e.m.f. is the voltage supplied to the external load; as we have seen, it acts in a direction so as to oppose what is causing it, that is, the motion of the rotor. In the motor the induced e.m.f. also acts in a direction so as to oppose what is causing it, rotor motion, and, in turn, what is causing the motion, that is the applied voltage. In a motor the induced e.m.f. is called the motor *back* e.m.f. The motor back e.m.f. is zero when the motor is not running and builds up as the motor speed builds up once the applied voltage is connected to the machine.

In both kinds of machine the armature has resistance and once current is established in the armature conductors there will be a potential difference across the armature. In the motor there must always be armature current for the motor to run so, when running, there is always an armature voltage drop. In the generator there will be an armature current only when an external load is connected. When the load is not connected no armature current flows and the generator output voltage is the induced e.m.f. When a load is connected there is an armature current and thus an armature voltage drop. The load voltage, that is, the generator terminal voltage is now the induced e.m.f. *reduced by* an amount equal to the armature voltage drop. The armature resistance is effectively the internal resistance of the supply in this case. In much the same way, any voltage source has an open circuit e.m.f. and a terminal p.d., the two being different depending upon the current drawn and the source internal resistance.

For a motor if we denote the applied voltage by V (volts), the back e.m.f. by E (volts) the armature current by I_a (amperes) and the armature resistance by R_a (ohms) the connecting equation is

$$V = E + I_a R_a$$

that is,

applied voltage = induced voltage + internal voltage drop

This equation may be rewritten as

$$E = V - I_a R_a$$

For the generator, using the same symbols E, I_a and R_a and denoting the generator terminal voltage by V (volts)

$$E = V + I_a R_a$$

that is,

> induced voltage = output voltage + armature volt drop

In both cases V represents the machine *terminal voltage, supplied to* the machine in the case of the motor and *supplied by* the machine in the case of the generator.

In general

$$E = V \pm I_a R_a$$

that is

> induced voltage = terminal voltage $\pm$ armature volt drop

This relationship is one of three important general equations for a machine.

Example 6.1   Calculate the e.m.f. generated by a d.c. machine providing 200 V at 15 A if the machine armature resistance is 0.5 Ω.

Armature volt drop = armature current $\times$ armature resistance

$$= 15 \times 0.5$$
$$= 7.5 \text{ V}$$

Machine terminal voltage (generator output) = 200 V

Thus, generated voltage $= 200 + 7.5$
$$= 207.5 \text{ V}$$

Example 6.2   The armature resistance of a motor is 0.6 Ω. Calculate the back e.m.f. of the motor when the armature current is 25 A and the supply voltage is 250 V d.c.

Armature volt drop $= 25 \times 0.6$
$$= 15 \text{ V}$$

Back e.m.f. = terminal voltage $-$ armature volt drop
$$= 250 - 15$$
$$= 235 \text{ V}$$

Induced voltage   The e.m.f. induced in a conductor situated in a changing magnetic field is directly proportional to the rate of change of flux. In a rotating machine the magnetic flux supplied by the pole pieces is constant and the change in the flux as it affects the conductor is obtained by movement of the rotor. The rate of change of the flux must therefore depend upon flux and the speed of rotation. Denoting flux per pole by ϕ (Wb) and speed of rotation by N (rev/min) we can write,

$$\text{induced voltage } E \propto \phi N$$

or

$$E = k_e \phi N$$

where k_e is a constant. The value of the constant depends upon the machine itself, it is affected by the number of conductors connected together and by the number of poles (the simple machine described so far has one pair of poles, more complex machines may have several pairs).

This equation is quite general and applies to both motors and generators equally.

Example 6.3   A d.c. generator running at 2000 rev/min and with a flux per pole of 0.15 Wb produces a generated voltage of 240 V. Calculate the generated voltage if the speed of the machine is increased to 2500 rev/min and the flux per pole is reduced to 0.1 WB.

The general equation connecting generated voltage E (volts), speed of rotation N (rev/min) and flux per pole ϕ (webers) is

$$E = k_e \phi N$$

where k_e is a machine constant. With this machine

$$240 = k_e \times 0.15 \times 2000$$

so that

$$k_e = \frac{240}{0.15 \times 2000}$$

$$= 0.8$$

and when $N = 2500$ rev/min and $0 = 0.1$ Wb, the generated voltage

$$E = 0.8 \times 0.1 \times 2500$$
$$= 200 \text{ V}$$

Torque

Both motors and generators when loaded produce torque. In the case of a motor this is the desired output, the inputs being an applied voltage and a magnetic field (or the current to produce a magnetic field). In the case of a generator the output is voltage, the inputs being torque (to turn the rotor) and a magnetic field. However, when current is drawn from the generator there exists a similar situation to the motor, namely armature current producing its own field and the main magnetic field, the reaction between them producing torque. In this case the torque acts *against* the input torque (in much the same way as the generated voltage or back e.m.f. of a motor acts against the applied voltage) and is overcome by it.

Torque is turning force, that is a force which causes or tends to cause rotation. It is measured by the product of the force and distance between the point of application of the force and the centre of rotation, see fig. 6.7. Its units are units of force × units of length, i.e. newton metres. In a rotating machine the distance between point of application (the conductor) and the centre of rotation (the rotor centre) is fixed for a particular machine, the force depends upon two variables; the flux per pole and the armature current (since the force is produced by interaction between two fields, that

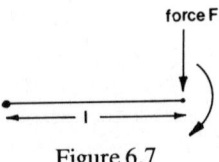

force F

Figure 6.7

produced by the poles and that produced by the armature current). Consequently, writing T (newton metres) for torque we can state

$$T \propto I_a \phi$$

where I_a (amperes) is the armature current and ϕ (webers) is the flux per pole.

The statement of proportionality may be written as an equation using a constant:

$$T = k_t I_a \phi$$

where the value of k_t depends upon the constructional details of the machine.

This equation is the third general equation applicable to both motors and generators (the other two are $E = V \pm$ internal volt drop and $E = k_e \phi N$).

Example 6.4   A d.c. motor taking an armature current of 30 A produces a torque of 60 Nm when the flux per pole is 0.2 Wb. Calculate the torque when the flux per pole is halved and the armature current is doubled.

Using the equation

$$T = k_t I_a \phi$$

we have

$$60 = k_t \times 30 \times 0.2$$

and

$$k_t = \frac{60}{30 \times 0.2}$$

$$= 10$$

so that when $I_a = 60$ A and $\phi = 0.1$ Wb
the torque

$$T = 10 \times 60 \times 0.1$$
$$= 60 \text{ Nm as before}$$

and we see that the effect of doubling the armature current exactly balances the effect of halving the flux per pole and the torque remains the same.

Mechanical power

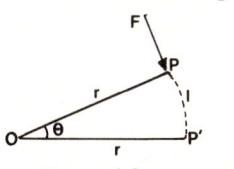

Figure 6.8

The mechanical power of a rotating machine (the input to a generator and the output from a motor) is related to the torque produced by the reaction between the main machine field and the field set up by the armature current. To determine the relationship a closer look at torque is required.

Consider the diagram of fig. 6.8. It shows an arm OP pivoted at O and rotating in a clockwise direction at N rev/min. The length of the arm is r metres. Suppose the arm moves to some new position OP′ in t seconds, the distance moved by point P in this time being l metres.

The tangential force, that is, the force acting at right angles on the arm (along a tangent to the arc PP′), which produces the movement from P to P′ is shown as F newtons.

By definition, torque is turning force and is the product of the force and the distance between the point of application of the force and the centre of rotation, in this case, Fr newton metres.

The work done by any force which moves its point of application is the product of the force and the distance moved, so that

$$\text{work done in moving from P to P′} = Fl \text{ newton metres}$$

and from geometry $l = r\Theta$ where Θ (rad) is the angle between OP and OP′, so that

$$\text{work done} = Fr\Theta$$

or, denoting torque by T (newton metres), work done $= T\Theta$ (joules).

This work is done in t seconds so that the power used in the movement from P to P′ (the rate of doing work) is given by

$$\text{power} = \frac{T\Theta}{t} \text{ joules/second or watts}$$

Now Θ/t is the angular velocity of the arm (rad/s) so that

$$\text{mechanical power} = \text{torque} \times \text{angular velocity}$$

The arm is rotating at N rev/min, that is $N/60$ rev/s, and, since one revolution is 2π rad, N revolutions is $2\pi N$ rad and $N/60$ rev/s is therefore $2\pi N/60$ rad/s, the angular velocity, hence

$$\text{mechanical power} = \text{torque} \times \frac{2\pi N}{60}$$

$$= \frac{2\pi NT}{60} \text{ watts}$$

The mechanical power as determined is the power produced by the machine torque. For the motor this is the power from which the output is obtained. For a generator this is the power which must be supplied by the prime mover (the means of making the generator rotor turn). With both machines there is also the question of losses to be considered.

Specific objectives

The expected learning outcome is that the student:
7.6   Defines efficiency of a machine as output/input.
7.7   Lists and explains the principal losses of machines.

Machine losses

Machines in general convert one kind of power into another. In electrical machines one kind of power is electrical and in rotating electrical machines the other kind of power is mechanical. Simply put, electric motors convert electrical power to mechanical power and generators do the reverse, convert mechanical power to electrical. In all machines there is a power input and a power output, the output always being smaller than the input owing to loss which takes place within the machine. As might be expected there are two main kinds of loss; electrical and mechanical. There is also a third kind of loss, in the magnetic circuit, which can be regarded more as an electrical loss rather than mechanical.

Electrical losses are due to conductive circuit resistance, the major loss occurring in the armature coils. Although armature resistance is usually very low, of the order of ohms (or less than an ohm), quite substantial currents flow in the armature and the power loss, which is equal to the product of resistance and the *square* of the current, can be relatively high. Armature losses are sometimes referred to as *copper losses* or I^2R *losses*. Other electrical losses include those at the brushes due to brush contact resistance and in the resistance of the field coils where these exist.

Mechanical losses include power lost in overcoming friction at the bearings of the rotating part of the machine and power lost in overcoming air resistance when the rotor turns. Modern bearings are extremely well designed to have minimum friction but it cannot be removed entirely. The 'air resistance' loss, usually called *windage*, is due to pushing air out of the way and setting up air currents within the machine as the rotor moves. Mechanical losses are often referred to as *friction and windage* losses.

Magnetic losses are due to hysteresis and eddy currents flowing in the magnetic circuit. As was described in *Electrical and Electronic Principles 2* the flux density, *B*, in a magnetic material does not change exactly in accordance with the magnetic field strength, *H*, so that when *H* is increased, for example, *B* will only increase so far before saturation occurs. As *H* is reduced the values of *B* corresponding to particular values of *H* are not the same as they were when *H* was being increased; there is a 'lagging' or *hysteresis* between *B* and *H*.

In magnetic materials being cycled through values of *B*, from saturation in one direction to saturation in another direction, the *B/H* graph is in the form of a loop called the hysteresis loop. Power is used in the process of changing the values of *B* and *H* as the material is cycled and the value of this power is proportional to the *area* of the hysteresis loop. This power is part of the magnetic loss. The remainder of the loss is due to the setting up of small currents, called eddy currents, which are induced in various random paths within the magnetic circuit (which is normally made of a material which is also an electrical conductor). These currents may be reduced by constructing the magnetic circuit of extremely thin slices called *laminations* but the currents are not completely removed and there is an associated power loss, as there is with any electric current flowing in a resistive material. The total magnetic losses (hysteresis and eddy current) are only a relatively small part of the total machine loss.

For a generator the input power comes from the machine which drives the generator, called the prime mover. The prime mover may be an engine of some kind (internal combustion etc.), a turbine driven by steam or water or may even be an electric motor. An electric motor driving a generator is used as one method of changing one kind or level of an electrical supply (that applied to the motor) to another kind or level of electrical supply (the output from the generator). The prime mover provides the input mechanical power to the generator, some of which is used in overcoming friction and windage, the remainder then being converted to electrical power in the generator armature. This total armature power is further reduced by armature copper losses and, if the machine provides its own excitation, by field losses (including magnetic losses and field winding resistance losses) before it is available at the output.

It was shown earlier that for a generator the induced e.m.f.

$$E = V + I_a R_a$$

where V is the output voltage, R_a the armature resistance and I_a the armature current.

If this equation is multiplied throughout by I_a, we obtain

$$E I_a = V I_a + I_a^2 R_a$$

The term $E I_a$ on the left hand side of this equation represents the total available power generated at the armature (that is, the input power from the prime mover reduced by the generator friction and windage loss).

The term $I_a^2 R_a$ represents the armature copper losses so that the remaining term, $V I_a$, is the available output power. This is further reduced by the field loss if the machine is self-excited.

The total electrical input power to an electric motor comes from the supply. If the field coils are connected across the same supply (shunt connection) the input power is reduced by the field loss, the remaining power being the input to the armature circuit. If the motor is series-connected, i.e. the field coils are in series with the armature, the whole of the input power to the motor is passed to the armature circuit, the field loss occurring *within* the armature circuit. The input power less the field loss is further reduced by the armature copper losses, the balance being available for conversion to mechanical power. There is a further loss due to friction and windage, the remainder of the mechanical power being the output mechanical power of the motor.

We can obtain an expression for the electrical power delivered to the armature as follows. Since

$$V = E + I_a R_a$$

where V is the applied voltage, E is the back e.m.f. and I_a, R_a the armature current and resistance, respectively, then

$$V I_a = E I_a + I_a^2 R_a$$

$V I_a$ is the electrical input power to the armature (the total motor input *less* the field loss for a shunt machine and the total motor input for series machine). $I_a^2 R_a$ represents the armature copper losses and $E I_a$ is the remainder of the armature power.

For a shunt machine this is the power converted to mechanical power; for a series machine this term is reduced by the field loss before conversion to mechanical power. The mechanical power is then reduced by friction and windage before being available at the output.

Efficiency Efficiency of a machine is the ratio of output power to input power. It may be expressed as per unit (p.u.) or as a percentage as follows

$$\text{Efficiency} = \frac{\text{output power}}{\text{input power}} \text{ p.u.}$$

or

$$\frac{\text{output power}}{\text{input power}} \times 100\%$$

Since

$$\text{output power} = \text{input power} - \text{losses}$$

then

$$\text{efficiency} = \frac{\text{input power} - \text{losses}}{\text{input power}} \text{ p.u.}$$

$$= 1 - \text{losses/input power p.u.}$$

$$\text{or } (1 - \text{losses/input power}) \times 100\%$$

The following examples should be studied carefully.

Example 6.5 A separately excited d.c. generator produces a generated e.m.f. of 220 V. The armature resistance is 0.6 Ω. When an armature current of 20 A is drawn from the machine, calculate:
  (a) the output voltage;
  (b) power loss in the armature conductors;
  (c) the output power from the machine.

(a) Using the symbols as before, since

$$E = V + I_a R_a$$

$$V = E - I_a R_a$$

$$= 220 - (20 \times 0.6)$$

$$= 208$$

The output voltage is 208 V.

(b) The power loss in the armature conductors (the copper loss) is equal to $I_a{}^2 R_a$

$$= 20^2 \times 0.6$$
$$= 240 \text{ W}$$

Armature power loss is 240 W.

(c) The available power from the armature

$$E I_a = V I_a + I_a{}^2 R_a$$

The term VI_a represents the remainder of the armature power after copper losses have been removed and is equal to

$$EI_a - I_a{}^2 R_a$$
$$= 220 \times 20 - 20^2 \times 0.6$$
$$= 4400 - 240$$
$$= 4160 \text{ W}$$

Since the machine is separately excited this power is not further reduced by the field loss and is available at the output.

Machine output power is 4160 W.

Example 6.6   A shunt connected d.c. motor has an armature resistance of 0.5 Ω and is supplied with 240 V at an armature current of 10 A.

The machine is running at 2000 rev/min and at this speed the friction and windage losses may be taken as 500 W. Calculate the output torque of the motor.

$$\text{Input power to armature} = VI_a$$
$$= 240 \times 10$$
$$= 2400 \text{ W}$$

$$\text{Armature copper losses} = I_a{}^2 R_a$$
$$= 10^2 \times 0.5$$
$$= 50 \text{ W}$$
$$\text{Friction and windage losses} = 500 \text{ W}$$

Electrical power available for conversion to mechanical power

$$= \text{armature input power} - \text{losses}$$
$$= 2400 - (500 + 50)$$
$$= 1850 \text{ W}$$

Now

$$\text{mechanical power} = \frac{2\pi \times \text{speed} \times \text{torque}}{60}$$

so that $1850 = 2\pi \times 2000 \times \text{torque}/60$ and

$$\text{torque} = \frac{1850 \times 60}{2\pi \times 2000} \text{ Nm}$$

$$= 8.83 \text{ Nm}$$

Output torque is 8.83 newton metres.

Notice that the field losses are not included in the 2400 W input supplied to the motor armature since this is a shunt connected machine. The total *motor* input in this case would be 2400 W *plus* the power supplied to the motor field coils (supply voltage × field coil current).

Example 6.7   A series connected d.c. motor is supplied with

400 V, the armature current being 8.5 A. The armature resistance is 0.4 Ω, the field coil resistance being 1.2 Ω. Friction and windage losses may be assumed to be 450 W and losses in the field magnetic circuit are equal to 20% of the field copper loss.

Input power to the armature (and to the motor since this is a series machine) $= 400 \times 8.5$
$$= 3400 \text{ W}$$
$$\text{Armature copper losses} = 8.5^2 \times 0.4$$
$$= 28.9 \text{ W}$$
$$\text{Field copper losses} = 8.5^2 \times 1.2$$
$$= 86.7 \text{ W}$$

(Notice the field current is also the armature current since the field coils are connected in series with the armature).

$$\text{Field magnetic losses} = 0.2 \times 86.7$$
$$= 17.34 \text{ W}$$
$$\text{Friction and windage loss} = 450 \text{ W}$$

$$\text{Total losses} = 450 + 17.34 + 86.7 + 28.9$$
$$= 582.94 \text{ W}$$

$$\text{efficiency} = 1 - \frac{\text{losses}}{\text{input}} \text{ (per unit)}$$

$$= 1 - \frac{582.94}{3400}$$

$$= 0.8285$$

The p.u. efficiency is 0.8285
(As a percentage this is 82.85%)

Specific objectives

The expected learning outcome is that the student:

7.8 *Distinguishes between shunt-wound and series-wound machines.*

7.9 *Sketches the load characteristics of the machines in 7.9 using the equation of 7.5.*

7.10 *Explains the machine characteristics in 7.9 using the equation of 7.5.*

7.12 *Identifies d.c. motors as variable speed machines.*

Armature reaction

When current flows in the armature of a d.c. machine, motor or generator, a magnetic field is set up by the armature as shown in fig. 6.9a.

This field acts at right angles to the main magnetic field, the resultant field being a vector combination of the two as shown in fig. 6.9b. This results in a weakening of the magnetic field on one side of the pole piece and a strengthening of the field on the other. The reluctance of a magnetic circuit depends upon flux density but it is not a linear relationship (the *B/H* curve) and the weakening effect is

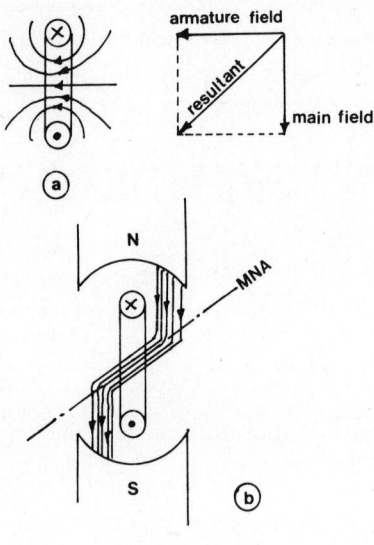

Figure 6.9

greater than the strengthening effect because of saturation and this results in an overall *reduction* in the strength of the magnetic field. The other result of the armature reaction is a shift in the position of the magnetic neutral axis (MNA) and therefore the best position for the commutator brushes. In a generator the brushes should be moved forwards (in the direction of rotation) and in a motor they should be moved backwards. In practice of course the actual best position and the position of the MNA depends upon how great is the effect of armature reaction and this in turn depends upon armature current (loading of the machine). If the machine loading is liable to fluctuate, repeated mechanical change of brush position is not convenient and in these cases additional poles, called interpoles, are provided to counteract the movement of the MNA. Interpoles are situated *between* the main poles and are usually magnetised by field coils connected in series with the armature. In this way the effect of the interpoles depends on the armature current as does the armature reaction which the interpoles are to counteract.

Armature reaction affects the characteristics of d.c. machines as discussed in the next section.

D.C. machine characteristics

The characteristics of d.c. machines are graphs of one variable quantity plotted aginst another, which quantities being determined by the type of machine. For a generator we are interested in generated e.m.f. and how this varies with excitation (i.e. the field current), with speed, and also with machine loading, i.e. the variation of the load current drawn from the machine armature. For a motor the important variables are speed and the output torque and how these vary with load. As the mechanical load on a motor is changed its speed tends to change and thus the back e.m.f. and also, as we shall see, the armature current, so we may measure the load on the machine by its armature current as with the generator.

Characteristics in general are affected by the field coil connections of the machine, series connection having different effects to those of parallel connection, a combination of the two types of connection (compound winding) combining, to some extent, the individual effects of series or parallel connections.

The following figures and associated comments should be studied carefully.

Separately excited generators

Fig. 6.10 shows the important characteristics of a generator having its field coils supplied by a separate source. The characteristics are:

(a) e.m.f./speed (constant field current);
(b) e.m.f./field current (constant speed);
(c) e.m.f. and terminal voltage/armature current.

As was stated earlier the generated e.m.f. is directly proportional to flux and to machine speed. In fig. 6.10a the graph is of the e.m.f. E plotted against values of speed N, with field current I_f and thus flux ϕ held constant.

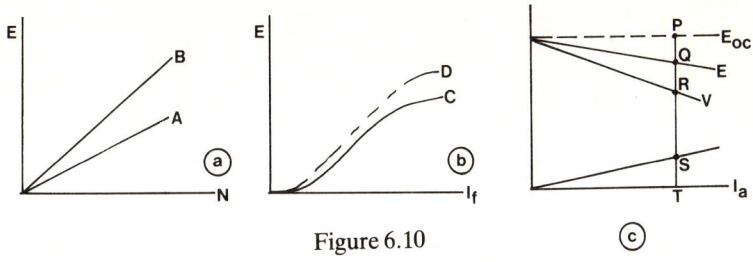

Figure 6.10

Since $E \propto \phi N$
and ϕ is constant
then $E \propto N$

and a straight line graph is obtained as in graph a. If the flux is increased to a new (constant) value the gradient of the graph rises as in graph b. Here the value of E for a particular value of N is greater than that in graph a for the same value of N.

If the speed is held constant

$$E \propto \phi$$

but flux ϕ is not directly proportional to field current I_f, the relationship being that of the B/H curve for the particular magnetic material used in the pole pieces.

Since the ϕ/I_f curve follows the B/H curve and $E \propto \phi$ the E/I_f graph will also follow the B/H curve as shown in graph c. This graph is drawn for constant speed. If the speed is increased the generated e.m.f. is greater at all points and the graph moves up to position D as shown.

The open circuit e.m.f. E_{oc} remains constant when the machine is run at constant speed and field current. On load, however, as armature current is increased, armature reaction *reduces* the effective flux as described earlier and the generated e.m.f. E/I_a graph droops as shown in fig. 6.10c. This figure also shows the graph of armature voltage drop $I_a R_a$ plotted against I_a (which is a linear rise as I_a is increased) and the graph of terminal voltage V plotted against I_a. As I_a rises $I_a R_a$ rises and since

$$V = E - I_a R_a$$

and E is falling due to armature reaction, V must also fall. At any point R on the graph of V/I_a

$$
\begin{aligned}
V &= RT \\
I_a R_a &= ST \\
\text{and} \quad E &= QT \\
\text{so that } QT &= RT + ST \text{ and } QR = ST
\end{aligned}
$$

Shunt excited generators   A shunt excited generator is one which provides its own excitation from the e.m.f. generated by the machine. The obvious immediate question that arises is how the machine can generate in order to provide the excitation when excitation is needed prior to generation. The answer lies in the residual magnetism retained by the field magnetic circuit. Unless there is some residual magnetism, generation cannot commence and full excitation and full generation will not be achieved.

Fig. 6.11 shows the three main characteristics of a shunt-connected generator. These are:

(a) generated e.m.f./speed;
(b) generated e.m.f./field current;
(c) generated e.m.f. and terminal voltage/armature current.

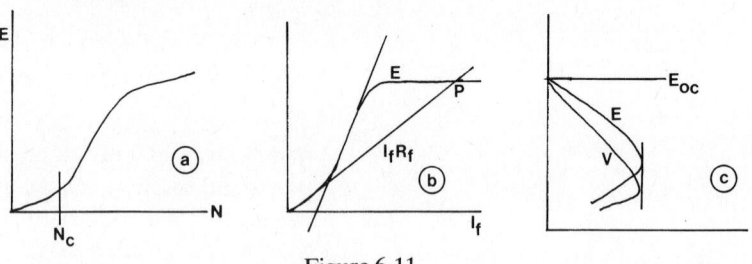

Figure 6.11

It will be seen that the e.m.f./speed graph shown in fig. 6.11a is not linear but to some extent seems to follow the familiar B/H shape.

Recalling that $E \propto \phi N$ it can be seen that on this occasion the flux ϕ cannot be held constant since it depends in turn on the induced e.m.f. so that we cannot obtain a $E \propto N$ relationship as with the separately excited machine. Examination of the graph shows that as the speed is increased little happens in the way of generation until at a particular speed, shown as N_c, the machine begins to generate in earnest. This speed is called the *critical speed* for the machine and depends directly on the field coil resistance as is explained below.

Fig. 6.11b shows a graph of generated e.m.f. E against field current I_f, and a graph of field coil voltage drop equal to $I_f R_f$, where R_f is the field circuit resistance, also against field current.

Point P represents the operating point of the machine and at this point the whole of the generated e.m.f. provides the field coil voltage drop $I_f R_f$. At lower values of field current (to the left of point P) which occur while the generator is being run up to the speed for which these graphs are drawn the generated e.m.f. is greater than the $I_f R_f$ drop, the difference being used in overcoming the back e.m.f. induced in the field coils due to the changing flux. At point P the magnetic circuit is saturated, the flux is constant and there is no induced e.m.f. in the field coils so that the whole of E is equal to the field coil voltage drop.

The slope of the $I_f R_f/I_f$ graph depends of course on the field circuit resistance, R_f. Commonly an additional variable resistor, called a *regulator* is connected in series with the field coils and R_f will also include the regulator resistance if there is one. If R_f is increased in value the slope of the $I_f R_f/I_f$ graph is increased and the operating point P will move along the E/I_f curve to the left and the available generated e.m.f. will be lower. If R_f is reduced in value point P moves to the right and the generated e.m.f. is higher. For all values of R_f the operating point occurs where the E/I_f curve and the $I_f R_f/I_f$

line intersect. If R_f is increased too much the $I_f R_f / I_f$ graph moves so far to the left it does not intersect with the E/I_f graph and the machine fails to excite and to generate. The reason for this, of course, is that for this position of the $I_f R_f$ graph the generated e.m.f. is insufficient to provide the necessary field coil voltage drop for any particular value of field current, let alone the extra voltage necessary to overcome the back e.m.f. in cases where the flux is still changing with field current.

For a particular value of R_f the $I_f R_f$ line is tangential to the E/I_f curve as shown in fig. 6.11b. This value of R_f, denoted R_{fc}, is called the *critical resistance* of the machine. For values of field current resistance below it the machine does not excite, for values above it the machine does excite. For each value of running speed there is a particular value of critical resistance, and vice versa, for each value of resistance there is a critical speed as shown in fig. 6.11a. The graphs of fig. 6.11a are for fixed field circuit resistance, the graphs of fig. 6.11b are for a particular fixed value of running speed.

The output characteristic of a shunt excited generator is shown in fig. 6.11c and appears somewhat unusual in that the E and V curves 'go back on themselves' once a particular value of armature (load) current has been reached. What is happening here is that as I_a is increased both E and V fall (as with the separately excited machine), the former due to armature reaction, the latter because it is equal to E less the armature voltage drop (which increases as I_a increases) and, at a particular value of I_a, E drops to a point where it can no longer provide the $I_f R_f$ voltage drop needed for full excitation. Any further attempt to increase output current I_a (by reducing the load resistance) merely causes E and thus V to drop further still and I_a in fact, falls off rather than increases. Taken to the limit, short circuit conditions apply and the generated e.m.f. falls to a very low value, due only to the residual magnetism since the field current is zero, the short circuit current being somewhat lower than the full load current. Although it is lower the machine may still sustain damage under these conditions.

Series excited generators

The particular point of importance with the series excited generator is that the field coil current I_f and armature current I_a are one and the same.

For a constant field (armature) current, the flux is constant and the generated e.m.f. is directly proportional to speed, as fig. 6.10a for the separately excited machine. The E/I_a graph is shown in fig. 6.12 and follows the familiar B/H shape since the armature current is also the field current. For any operating point P, as shown, the line OP has a slope equal to the *combined* load and field circuit resistance, the whole of the generated e.m.f. being used to provide the voltage drop across it. The line OP is, of course, the graph of the product of I_f (I_a) and the total load and field circuit resistance plotted against I_f (I_a) and if the combined resistance has a value such that this graph lies to the left of the E/I_a curve the machine does not excite as with the shunt machine. There is therefore a critical *load*

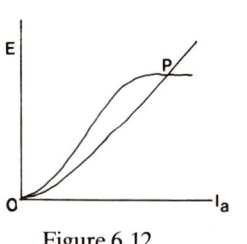

Figure 6.12

resistance wth this machine for a particular value of field circuit resistance.

The shape of the load characteristic, giving as it does wide variations in generated voltage for different values of load current, renders the machine unsuitable for most applications.

Compound wound generators

The 'droop' of the load characteristics of shunt connected machines (fig. 6.11c) may be reduced by the addition of series field coils bringing in a rising value of E as I_a rises as shown in fig. 6.12. A generator containing both series and parallel field coils is called *compound wound* and the overall shape of the load characteristic is determined by the number of series coils and how they are connected. See fig. 6.13.

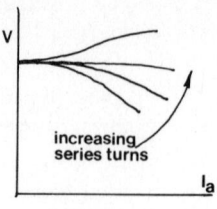

Figure 6.13

Shunt motors

The output torque of a motor is directly proportional to both flux and armature current as was discussed earlier. Using the usual symbols

$$T \propto \phi I_a$$

In a shunt motor the field coils are connected in parallel with the armature so that the field current and thus the flux are substantially constant (neglecting for the moment the effects of armature reaction which weakens the flux as armature current is increased). We can write then that

$$T \propto I_a$$

which would give a straight line graph when T is plotted against I_a. The effects of armature reaction tend to make the graph shape depart slightly from a straight line in practice, see fig. 6.14.

The motor back e.m.f. E is connected to the applied voltage V, the armature current I_a and the armature resistance R_a by the equation

$$E = V - I_a R_a$$

and is directly proportional to flux ϕ and speed N

$$E \propto N \phi$$

so that the motor speed

$$N \propto \frac{E}{\phi}$$

and thus

$$N \propto \frac{V - I_a R_a}{\phi}$$

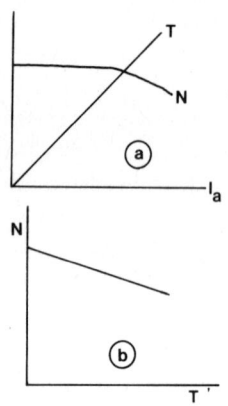

Figure 6.14

As the armature current is increased the term $(V - I_a R_a)$ is reduced in value and so is the flux ϕ due to armature reaction. The effect on the overall expression

$$\frac{V - I_a R_a}{\phi}$$

depends upon the relative sizes of these reductions and in practice the numerator $V - I_a R_a$ is reduced slightly more than the flux ϕ for a particular increase in I_a. The difference is slight so that the shunt motor may be considered to be a substantially constant speed machine. See fig. 6.14. The speed/torque curve, which may be deduced from the other curves is also shown in the figure.

Series motors

In the series machine the field coils are connected in series with the armature so that the field current and armature current are one and the same. The flux is produced by the field current and if the machine is not saturated is approximately proportional to it (the non-saturation part of the B/H curve).

Consequently the statement of proportionality

$$T \propto \phi I_a$$

becomes

$$T \propto I_a$$

since

$$\phi \propto I_a \text{ approximately}$$

After saturation ϕ can no longer be increased by increasing I_a and is constant so that $T \propto I_a$, which is a linear relationship. This gives a torque/armature current graph of the shape shown in fig. 6.15, the distinct change in shape occurring as the machine becomes saturated.

Earlier we saw that motor speed

$$N \propto \frac{V - I_a R_a}{\phi}$$

(which is true however the motor is connected). In a series machine the flux ϕ is approximately proportional to the armature current prior to saturation and as I_a rises so does ϕ. Neglecting the armature voltage drop, $I_a R_a$, for the moment we have the approximate relationship

$$N \propto \frac{V}{\phi}$$

which would produce a rectangular hyperbole when speed is plotted against armature current. In practice this shape is slightly modified by the increasing armature voltage drop, which tends to push the curve down towards the armature current axis. As saturation occurs the speed tends to become constant at a relatively low value. The speed/armature current graph is shown in fig. 6.15. Note that as I_a is reduced the speed rises to a high value so that on no load the speed becomes dangerously high. A series motor should not therefore be disconnected from its load. Fig. 6.15 also shows the speed/torque curve which may be deduced from the other curves.

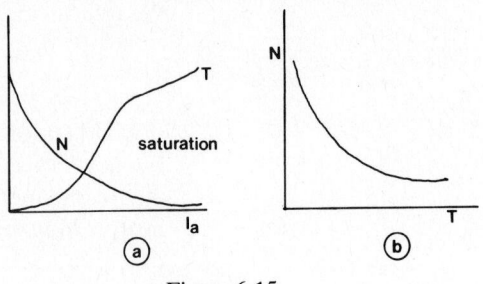

Figure 6.15

Compound motors   A compound motor has part of the field circuit in series with the armature and part in parallel. The characteristics are thus a compromise between those of the series machine and those of the parallel machine, the exact shape being determined by the relative effects of the series and parallel field coils. Typical curves are shown in fig. 6.16, 'cumulative' meaning that the two sets of coils aid one another, 'differential' meaning that they act in opposition.

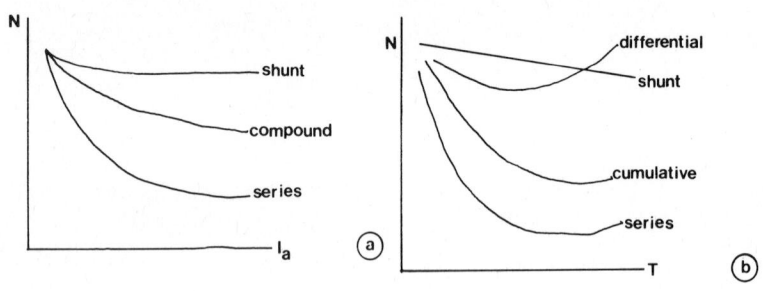

Figure 6.16

Specific objectives   The expected learning outcome is that the student:
7.11   *Explains the need for a d.c. motor starter.*

D.C. motor starters   The basic motor equation relating back e.m.f. to other variables is

$$E = V - I_a R_a$$

and also

$$E \propto N \phi$$

On starting the motor, at the instant of applying the supply voltage the motor speed is zero and so is the back e.m.f. so that

$$V = I_a R_a$$

The armature resistance is extremely low and a very high current would flow in the armature under these conditions. To prevent such a current and possible damage to the motor, specially designed motor starters are used, a typical circuit arrangement being shown in fig. 6.17.

The motor starter places a number of resistors in series with the

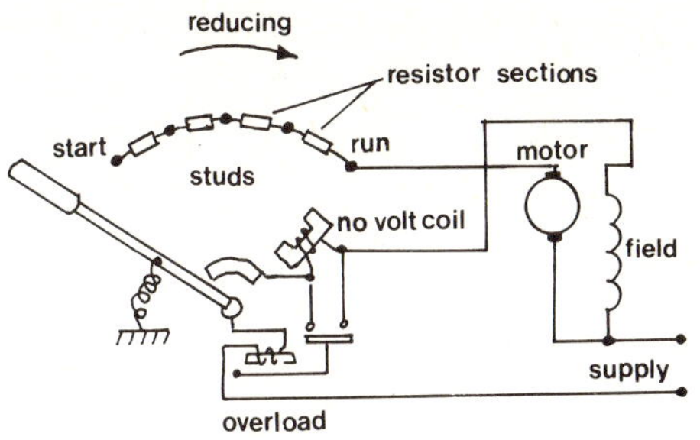

Figure 6.17

motor armature as the starter handle is moved from the rest position. This additional resistance reduces the initial armature current to a safe value while the motor speed builds up allowing the back e.m.f. to do likewise. As the handle is moved towards the 'run' position the additional resistance is reduced progressively to zero by the sliding contact of the handle touching a series of resistor studs in turn. The handle is held in the 'run' position by the 'no volt' coil, as shown, so that should the supply be interrupted the handle is released and the starting procedure must be recommenced. An additional overload coil mechanism short circuits the no volt coil if the load current is excessive, allowing the handle to return to the rest position and breaking the motor circuit. D.C. motor starters should be operated smoothly, neither too slowly nor too quickly, for most effective and reliable starting.

Specific objectives

The expected learning outcome is that the student:

7.14 *Describes how a symmetrical polyphase supply applied to a symmetrical polyphase winding produces a resultant travelling field of uniform magnitude.*

7.15 *Explains how a rotating field induces currents in a conducting rotor and produces torque in an induction motor.*

7.16 *Defines induction motor slip.*

7.17 *Explains why slip increases with load.*

The induction motor

The induction motor is one of the most widely used machines for a variety of industrial and other purposes. Its principal advantages are relative cheapness, reliability and efficiency and its robust construction makes it extremely useful for application in hostile environments.

The principle of the induction motor is that a rotating magnetic field is produced by the stator coils and this moving field induces currents in the rotor. The induced currents set up their own magnetic field which interacts with the rotating field to produce rotor movement. No supply is taken to the rotor since there is no need and this clearly makes for simplicity of construction.

There are two types of induction motor, three-phase a.c. and single-phase a.c. We shall be primarily concerned with the first of these.

Production of rotating field

Fig. 6.18 should be studied carefully. It shows the basic layout of stator coils, waveforms of the three-phase supply and phasor diagrams showing component and resultant m.m.f.s within the machine.

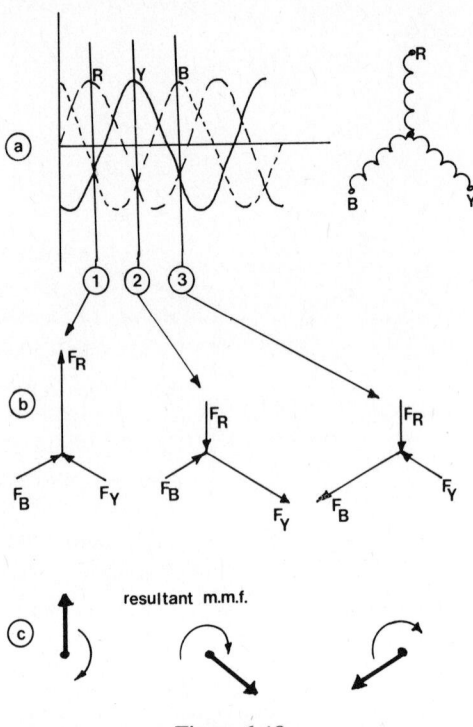

Figure 6.18

For purposes of explanation it is assumed that a current shown as positive in the waveform diagram produces an m.m.f. acting outwards from the centre of the machine along the axis of the coil producing it, the coils being mounted along radii. A negative current produces an m.m.f. acting inwards along the radius on which the coil lies. Three different instants of time are taken, shown as 1, 2 and 3, and m.m.f. diagrams are shown for each instant.

Fig. 6.18a shows the three phase waveforms, R, Y and B in sequence, 120° displaced from each other, fig. 6.18b shows the m.m.f. diagrams at time instants 1, 2 and 3 and also the coil layout and fig. 6.18c shows the resultant m.m.f.

At instant 1 the m.m.f. produced by coil R, shown as F_R, shown as F_R, acts outwards. The m.m.f.s produced by coils Y and B, shown as F_Y and F_B, respectively, act inwards. If the coils are identical in each phase at this point $F_R = 2F_Y$ and $F_Y = F_B$ since, from the diagram, $I_R = 2I_Y$ and $I_Y = I_B$.

F_Y may be considered to have two components:

a horizontal component $F_Y \cos 30$ acting to the left

and

a vertical component $F_Y \cos 60$ acting upwards.

Similarly, F_B may be considered to have two components:

a horizontal component $F_B \cos 30$ acting to the right

and

a vertical component $F_B \cos 60$ acting upwards.

The horizontal components cancel since they are equal and opposite, the vertical components add together to give

$$F_Y \cos 60 + F_B \cos 60 \text{ acting upwards}$$
$$\text{i.e. } 0.5\,F_Y + 0.5\,F_B$$
$$\text{or } F_Y \text{ or } F_B \text{ (since } F_Y = F_B\text{)}$$

which adds directly to F_R to give $1.5\,F_R$ acting vertically upwards (since $F_Y = F_B = 0.5\,F_R$).

The resultant m.m.f. at this instant of time is therefore $1.5 \times$ maximum m.m.f. produced per phase and acts vertically upwards.

At instant 2, F_Y now a maximum, acts outwards, F_R and F_B, equal to $0.5\,F_Y$, act inwards and again by resolution the total resultant m.m.f. is $1.5 \times$ maximum m.m.f. per phase *but the line of action has moved* to that of F_Y, i.e. acting downwards at an angle of $60°$ to the right of the vertical as shown in fig. 6.18c.

At instant 3, F_B is at a maximum acting outwards and the resultant m.m.f. is again $1.5 \times$ maximum m.m.f. per phase, on this occasion acting along the line of action F_B, acting downwards at an angle of $60°$ to the left of the vertical.

We have taken only three instants of time and examination of waveforms and m.m.f.s appears to show that a constant value m.m.f. acts, and its line of action rotates, in a clockwise direction for the phase connection to the coils R, Y, B shown in the figure. A large number of examples taken at many different instants of time confirms that this is so. The three coils produce in effect 'rotating pole pieces' which set up a magnetic field which acts through the rotor centre and which is itself rotating. The action of this field makes the rotor rotate also as explained below.

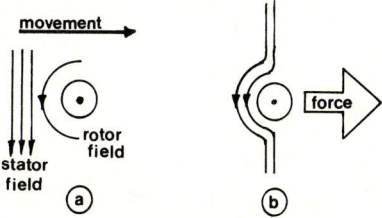

Figure 6.19

When the magnetic field set up by the three-phase supply applied to the stator moves past a rotor conductor an e.m.f. is induced in the conductor. If the conductor forms part of a closed circuit an electric current flows which establishes its own magnetic field as shown in fig. 6.19a. The fields interact and produce force as shown in fig. 6.19b and the rotor conductor, being able to move, does so. This is the basic principle of the three-phase induction motor.

In the single-phase induction motor, which will not be considered in any detail, the single-phase supply is divided into two components acting in quadrature with each other (i.e. phase displaced by $\pi/2$ rad or a quarter of a cycle) on starting, which produces a rotating field in a similar manner to that considered above. Once started it is found that the phase displacement is no longer required and the motor continues to run. Single-phase induction motors are used extensively in domestic and small industrial equipment.

Slip   In order for induction in the rotor conductors to take place the magnetic field produced by the stator must move past the rotor conductors, cutting them. If the rotor conductors move at the same speed as the magnetic field, cutting does not take place and induction (which causes the motion) does not occur. The rotor therefore moves at a slower speed than that of the magnetic field. Any tendency for the rotor speed to increase reduces the difference between the rotor speed and the speed of the magnetic field. This, in turn, reduces the rate of change of flux and the induced rotor voltage and current. Reduction in rotor current produces a reduction in turning force, that is, the motor torque, and the initial rise in rotor speed is offset by the slowing of the rotor due to this torque reduction.

The speed of the rotating magnetic field is called the motor *synchronous speed* and is a function of supply frequency. The difference between the synchronous speed and the rotor speed is called the *slip* of the motor. Denoting synchronous speed by N_s (rev/min) and motor speed by N (rev/min) the fractional slip of the motor is defined as

$$\frac{N_s - N}{N} \quad \text{per unit (p.u.)}$$

This may be expressed as a percentage by multiplying by 100.

As the induction motor is more heavily loaded, i.e. the rotor is presented with greater resistive force to overcome, the machine rotor tends to slow and the motor slip increases. This increases the induced currents in the rotor thereby increasing the torque to cope with the increased load.

Summary   Electrical machines are energy converters. Motors convert electrical energy to mechanical energy; generators convert mechanical energy to electrical energy. Both kinds of machine have within them a magnetic field system, usually established by field coils through which current is passed, and an armature made up of a number of interconnected conductors. The interaction between the machine field and the armature produces the end result which in a motor is force, usually a turning force or torque, and in a generator is an output voltage. Both kinds of machine usually have a part which rotates, called the rotor, and a part which does not, called the stator.

A motor is supplied with field current and armature voltage and produces torque. A generator is supplied with field current and torque and produces armature voltage. However, an e.m.f. is induced and a torque set up in both kinds of machine. In a motor the induced e.m.f. opposes part of the applied voltage and is called a back e.m.f. In a generator the torque established by the machine opposes part of the input torque.

The relevant equations which apply equally to both kinds of machine are

$$E = Blv$$

and

$$F = BlI$$

where E is the induced e.m.f. and F the force set up, B is the machine flux density, l the length of a conductor, v the velocity of the conductor relative to the field and I the current in the conductor. These equations may be developed to give the standard machine equations

$$R = k_e \phi N$$
$$T = k_t \phi I_a$$

where
E is the generated voltage
ϕ the machine flux
N the machine speed
T the machine torque
I_a the armature current
k_e, k_t are constants for a particular machine.

D.C. machines are supplied with or supply direct current. Connections to the armature, when this is situated on the rotor, are made by brushes pressing on copper segments, each one attached to one or more conductors, the arrangement of copper segments being called a commutator. The commutator acts as a mechanical rectifier.

In both kinds of machine there is a generated e.m.f. and a supply voltage, supplied to a motor and by a generator, and an internal voltage drop within the machine due to conductor resistance. Denoting generated e.m.f. by E and supply voltage by V the general equation is

$$E = V \pm \text{internal volt drop}$$

For a motor

E (back e.m.f.) $= V$ (supply voltage) $-$ internal voltage drop

For a generator

E (generated e.m.f.) $= V$ (output voltage) $+$ internal voltage drop.

The per unit (p.u.) efficiency of a machine is defined as

$$\frac{\text{output power}}{\text{input power}}$$

the percentage efficiency being p.u. efficiency $\times$ 100%. Since output power = input power − losses this expression may also be written as (1 − losses/input power). The principal power losses in a machine are those due to armature resistance, called (armature) copper losses and those due to friction and windage. In addition there are magnetic losses due to hysteresis of and eddy currents in the magnetic circuit within the machine. Losses due to the resistance of field coils, when these are present, must also be taken into account depending upon how the field coils are connected. Field coils may be provided with a separate supply, when the machine is described as separately excited, or may be connected in parallel or series with the armature, when the machine is termed shunt or series excited respectively.

Machine characteristics are graphs of important variable quantities plotted against others. For a generator the quantities of interest are generated and output voltage plotted against field current or armature current or speed. For a motor the important quantities are torque and speed plotted against armature current and speed plotted against torque. Characteristics show that the shunt and separately excited generators are relatively constant voltage machines, at any particular speed, the series generator having a much more variable speed/armature current graph. In addition, for the latter machine at any particular speed there is a critical value of the field circuit resistance above which the machine will not start to generate. This resistance is called the critical resistance. For any particular field circuit resistance, there is, similarly, a critical speed. The shunt motor is essentially a constant speed machine, the series motor having a characteristic such that speed falls dramatically as speed is increased. Series machines may run at dangerously high speeds when the mechanical load is disconnected. For both series and shunt machines torque rises with speed, for the latter relatively linearly and for the former initially roughly following a B/H curve shape then, following magnetic saturation, approximately linearly as the shunt machine. Both generators and motors may have a mix of series and parallel field coils and are then called compound wound machines. The characteristics of compound wound machines are a combination of those of series and shunt machines, the exact shape depending upon the relative effect of each kind of winding.

The value of the back e.m.f. in a d.c. motor depends directly on the running speed. On starting this is zero and the whole of the supply voltage applied to the low resistance armature circuit may cause extremely high currents. To prevent this, specially designed motor starters are used, in which the total resistance is gradually reduced as the machine is started.

The a.c. induction motor is an extremely versatile machine, widely used and is robust, reliable and efficient. The basic principle is the production of a rotating magnetic field by specially arranged stator coils, the field then cutting the rotor conductors and inducing voltage and current. The rotor currents set up their own magnetic fields which interact with the main rotating field producing torque and motion. The supply to an induction motor is commonly three-phase a.c. but may be single-phase provided that the machine is specially wound and contains a starter circuit which divides the supply voltage into two quadrature components. The rotating field speed is called the synchronous speed of the machine and the difference between the running speed and the synchronous speed is called the slip which may be expressed as per unit or as a percentage. Machine slip increases as the load on the induction motor is increased.

EXERCISE 6.1   (The effects of armature reaction may be neglected in the following questions.)

1. A d.c. shunt machine has an armature resistance of 0.1 Ω and field coil resistance of 120 Ω. As a generator running at 400 rev/min it delivers 120 kW at a constant voltage of 375 V. Calculate the speed of the machine when supplied with 375 V and 120 kW input power.

2. Calculate the torque developed by a 600 V d.c. shunt motor running at 1000 rev/min if the armature resistance is 0.4 Ω and armature current is 50 A.

3. A d.c. shunt motor running at 1500 rev/min takes 30 A armature current from a 440 V supply. When 240 V is supplied the armature current falls to 20 A and the flux is reduced by 25% of the initial value. Calculate the speed at this value of applied voltage if the armature resistance is 0.5 Ω.

4. A d.c. generator running at 1500 rev/min and with a flux per pole of 0.12 Wb produces a generated voltage of 300 V. When the flux per pole is reduced and the speed is increased to 3000 rev/min the generated voltage is 260 V. Calculate the new flux/pole.

5. A separately excited d.c. generator produces a terminal voltage of 200 V when 25 A is drawn from the armature. The armature resistance is 0.4 Ω. Calculate:
   (a) the generator voltage;
   (b) the output power from the machine.

6. The output torque of a 250 V shunt connected d.c. motor is 12 Nm at an armature current of 15 A. Friction and windage losses are 450 W, the armature resistance being 0.6 Ω. Calculate:
   (a) the total losses in the machine;
   (b) the speed of the motor.

7. A shunt connected d.c. motor has a p.u. efficiency of 0.85 and an input power of 4 kW. The armature copper loss is 150 W, the field copper loss being 35 W. Calculate the friction and windage loss of the machine.

8. A shunt wound d.c. generator providing its own excitation generates 450 V at an armature current of 30 A. The armature and field coil resistances are 0.5 Ω and 50 Ω, respectively, the friction and windage loss being 400 W. Calculate the per unit efficiency of the machine.

Possible marks

SELF-ASSESSMENT EXERCISE 6   1. State the equation relating the e.m.f. induced across a conductor of length *l* metres, moving at *v* metres/second through a magnetic field of flux density *B* tesla.   (3)

2. State the equation relating the force acting on a conductor of length *l* metres carrying a current of *I* amperes and situated in a magnetic field of flux density of *B* tesla.   (3)

3. State the general d.c. machine equation relating induced e.m.f. E volts, terminal voltage V volts, armature current I_a amperes and armature resistance R_a ohms. (3)

4. Define per unit efficiency for a d.c. machine and list the power losses in the machine. (3)

5. State the equation defining per unit slip for a three-phase induction motor. (3)

6. A d.c. generator provides 420 V at an armature current of 10 A, the generated voltage being 450 V. What will be the terminal voltage at an armature current of 20 A, the generated voltage now being 440 V? (5)

7. A 220 V d.c. motor with an armature resistance of 0.8 Ω has a back e.m.f. of 180 V. Calculate the armature copper losses. (5)

8. A d.c. generator running at 1000 rev/min and with a flux per pole of 0.1 Wb produces a generated voltage of 400 V. Calculate the generated voltage if the speed of the machine is increased to 1500 rev/min and the flux per pole is doubled. (5)

9. A separately excited d.c. generator having an armature resistance of 0.4 Ω provides a terminal voltage of 200 V at a load current of 20 A. The armature copper loss is 250 W. Calculate:
   (a) the no-load voltage;
   (b) the machine output power;
   (c) the input power assuming a p.u. efficiency of 0.82. (14)

10. A 240 V shunt connected d.c. motor is running at 1000 rev/min and takes an armature current of 20 A. The armature copper loss is 200 W, friction and windage losses being 400 W. Calculate the:
   (a) armature resistance;
   (b) motor back e.m.f.;
   (c) motor output torque. (14)

11. Explain what is meant by the following terms in connection with d.c. machines:
   (a) armature copper loss;
   (b) commutator;
   (c) magnetic neutral axis;
   (d) armature reaction;
   (e) interpoles;
   (f) per unit efficiency;
   (g) friction and windage loss. (14)

12. A shunt machine driven at 2000 rev/min generates a direct voltage of 250 V at 40 A. If the machine runs as a motor with the same terminal voltage and an armature current of 50 A calculate the speed. Armature copper losses when the machine is being run as a motor are 1000 W. Neglect the effects of armature reaction. (14)

13. With the aid of current waveform and other relevant diagrams describe how a rotating magnetic field is set up and used to supply motive power in a three-phase induction motor. What is meant by per-unit slip of an induction motor? (14)

Answers

1. 337.6 rev/min

2. 276.9 Nm

3. 1082.4 rev/min

4. 0.052 Wb

5. (a) 210 V (b) 5 kW

6. (a) 585 W (b) 2518.6 rev/min

7. 415 W

8. 0.647

SELF-ASSESSMENT EXERCISE 6

1. $E = Blv$ (3)

2. $F = BlI$ (3)

3. $E = V \pm I_a R_a$ (3)

4. p.u. efficiency = $1 -$ losses/input (1)
Losses include copper losses, friction and windage losses and magnetic circuit losses. (2)

5.
$$\text{p.u. slip} = \frac{N_s - N}{N_s}$$

where N_s is synchronous speed and N is running speed. (3)

6. $450 = 420 + 10R_a$ (1½)
Therefore $R_a = 3\,\Omega$ (1)
$440 = V + 20 \times 3$ (1½)
Therefore $V = 380$ V (1)

7. $180 = 220 - I_a \times 0.8$ (1½)
Therefore $I_a = 50$ A (1)
Armature copper losses $= 50^2 \times 0.8$ (1½)
    $= 2$ kW (1)

8. $E = k_e N \phi$ (1)
$400 = k_e \times 1000 \times 0.1$ (1)
$E = k_e \times 1500 \times 0.2$ (1)
Therefore $E = 400 \times \dfrac{1500}{1000} \times \dfrac{0.2}{0.1}$ (1)

    $= 1200$ V (1)

9. (a) No load voltage $= 200 + 20 \times 0.4$
    $= 208$ V (3)

(b) Generated power $= 208 \times 20$
    $= 4160$ W (3)
Armature copper loss $= 20^2 \times 0.4$
    $= 160$ W (2)
Output power $= 4160 - 160$
    $= 4000$ W (2)

(c)
$$\text{Input power} = \frac{4000}{0.82} = 4878 \text{ W} \quad (4)$$

10. (a) Armature copper loss $= 200$ W
Therefore $20^2 \times R_a = 200$
and $R_a = 0.5\,\Omega$ (3)

(b) $E = 240 - 20 \times 0.5$
    $= 230$ V (3)

(c) Input power to armature
    $= 240 \times 20$
    $= 4800$ W (2)
Losses $= 200 + 400$
    $= 600$ W (2)
Output power $= 4800 - 600$
    $= 4200$ (1)

Therefore $4200 = 2\pi \times 1000 \times T/60$ (2)

and

$$T = \frac{4200 \times 60}{2\pi \times 1000} = 40.1\,\text{Nm} \tag{1}$$

11. Answers as text.   (2 per answer)

12. Armature copper losses = 1000 W at an armature current of 50 A.

Therefore

$$R_a = \frac{1000}{2500} = 0.4\,\Omega \tag{2}$$

Therefore generator terminal voltage = $250 - 40 \times 0.4$
$$= 234\,\text{V} \tag{3}$$

Motor terminal voltage = 234 V
Motor back e.m.f. = $234 - 50 \times 0.4$ (3)
$$= 214\,\text{V}$$

Using $E = k_e N \phi$

generator $250 = k_e \times 2000 \times \phi$ (2)
motor $214 = k_e \times N \times \phi$ (2)

(ϕ remains the same neglecting armature reaction since field coil voltage, i.e. terminal voltage, is the same) so that

$$N = \frac{2000 \times 214}{250} = 1712\,\text{rev/min} \tag{2}$$

13. As text.   (waveform diagram 3)
(m.m.f. diagram 6)
(explanation of m.m.f. resultant rotation 3)
(definition of p.u. slip 2)

7 Control and measurement

Topic areas: G, H and I

In this chapter the remainder of the standard unit Electrical and Electronic Principles 3 will be considered. The three topic areas concerned with magnetic amplifiers, various electronic components and devices and the use of measuring instruments may be conveniently and logically grouped together under the general heading 'Control and measurement' since the magnetic amplifier and the various electronic components considered are used primarily in the control of power and its application to plant, machinery and other electrical loads, the remainder of the chapter being concerned with the measurement of electrical characteristics of these and other circuits.

General objective *The expected learning outcome is that the student understands the operation of the magnetic amplifier as a variable impedance device.*

Specific objectives *The expected learning outcome is that the student:*
10.1   *States that the inductance of an iron-cored coil depends upon the permeability of the iron.*
10.2   *States the relationship between permeability of iron and the magnetic force.*
10.3   *Describes how the impedance of a simple transductor may be varied by the control coil current.*
10.4   *Sketches the circuits of simple magnetic amplifiers with balanced input and output circuits.*
10.5   *Explains the purpose and method of application of bias in magnetic amplifiers.*
10.6   *Explains the purpose and method of application of feedback magnetic amplifiers.*
10.7   *Lists and describes the advantages and disadvantages of magnetic amplifiers in comparison with semiconductor amplifiers.*

Magnetic amplifiers   The magnetic amplifier has as its principal advantage the ability to control high levels of power by the application of small signals, hence the name 'amplifier'. It is essentially an accurately controlled switch having a continuously variable operating point in a similar manner to that of the thyristor and triac to be examined later. The magnetic amplifier depends for its operation on the variation of the inductance of iron cored coils due to the level of magnetic flux in the coil core; to gain an appreciation of operation it is first necessary to revise the basic principles of self-inductance.

It will be recalled that self-inductance of a conductor is the property of setting up an induced e.m.f. which opposes change in the current flowing in the conductor. The e.m.f., called a back e.m.f. because of its opposition to change, is induced by the variation in the magnetic flux due to the conductor current caused when the current changes its level.

The magnitude of the back e.m.f. is directly proportional to the rate of change of the conductor current, the constant of proportionality being the self-inductance of the coil.

Symbolically, the back e.m.f.

$$E = -L\frac{di}{dt}$$

where L is the self-inductance (henrys) and di/dt is the rate of change of current with time (amperes/second).

The self-inductance of a particular conductor depends upon:

(a) the physical dimensions of the coil core (cross-sectional area and length);

(b) the number of turns of the conductor (winding the conductor in the form of a coil increases the magnetic flux per unit current);

(c) the magnetic properties of the core, specifically, the absolute permeability μ.

The relationship mathematically (derived on pp. 67–68 *Electrical and Electronic Principles 2*) is

$$L = N^2\frac{\mu A}{l}$$

where

N is the number of turns
μ is the core absolute permeability (henrys/metre)
A is the core cross-sectional area (metres^2)
l is the coil conductor length (metres)

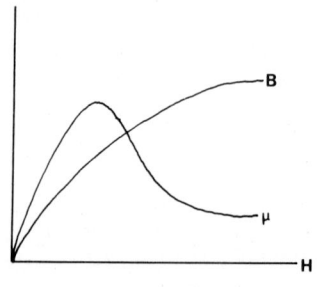

Figure 7.1

Absolute permeability of a magnetic material is the product of the permeability of free space μ_0, which is constant and equal to $4\pi \times 10^{-7}$ H/m, and the relative permeability of the material which depends specifically on the state of magnetisation of the core. Absolute permeability is in fact the ratio of magnetic flux density B (tesla) to magnetising force H (ampere-turns/metre) and may be calculated from the graph of B plotted against H, the B/H curve. Fig. 7.1 shows a typical graph of B plotted against H and also of μ plotted against H (the graph of μ_r/H being of similar shape since $\mu_r = \mu/\mu_0$ and μ_0 being constant). As is seen there the coil core permeability varies and *so therefore will the coil self-inductance* in a similar manner as the value of the magnetising force H and thus the current (which establishes the magnetising force) is varied.

Consider now an idealised linear B/H graph as shown in fig. 7.2. From the origin to point X the flux density increases linearly with magnetising force and so the value of B/H and therefore the permeability is constant. Over this region therefore the coil self-inductance is also constant. This also applies over the range from the origin to point Y where both *B* and *H* are reversed in direction.

To the right of point X, however, the idealised curve is a straight line parallel to the *H* axis and increasing *H* has no effect on *B*. The core is saturated, the flux density is constant and the ratio *B/H* and therefore the absolute permeability rapidly become zero as *H* is

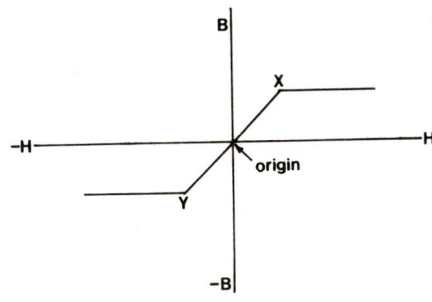

Figure 7.2

increased. The value of the coil inductance in this region, following the value of absolute permeability, also falls rapidly to zero as the operating point moves further to the right of point X along the *B/H* graph. Similar considerations apply to the left of point Y in the lower part of the graph where the direction of both *B* and *H* is reversed. The magnetic amplifier works on this principle of driving a magnetic core into and out of saturation forcing the permeability and thus the self-inductance to zero. We shall examine the effects of this in circuit in a moment. Practically of course, it is not possible to attain an absolutely linear *B/H* relationship as shown in the idealised curve of fig. 7.2. Special core materials have been developed and are used in magnetic amplifiers to obtain a situation corresponding as closely as possible to the ideal.

The saturable reactor

The circuit of fig. 7.3 shows a core having two windings connected so that one winding is in series with a load and an a.c. supply, the other winding being provided with a separate d.c. supply.

The winding in series with the load is called the *load winding*, the other winding is called the *control winding* and the device as a whole is called a *saturable reactor*.

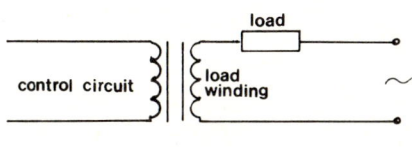

Figure 7.3

Provided that the a.c. supply on its own does not drive the core into saturation and the load winding impedance in this condition is much greater than that of the load, the bulk of the a.c. supply voltage will be across the load winding when no current flows in the control winding. If now a current is allowed to flow in the control winding an additional flux is established in the core. The a.c. supply in the load circuit establishes a magnetic flux first in one direction and then the other. When the direction of this flux acts so as to assist the flux caused by the control winding, the core may be driven into saturation. If this occurs the core permeability and thus the winding coil inductance is driven to zero and during the period of saturation the bulk of the a.c. supply voltage now appears across the load and a substantial load current flows. The average value of the load current depends directly on the time per half cycle that the core is driven into saturation. The point at which the core enters and leaves saturation in turn depends upon the flux already established by the control winding current itself. There is thus a direct relationship between average load current (output) and d.c. control current (input) and by correct design very small values of control current can control large values of load current and thus load power. The circuit as shown is not in fact practicable for, as may have been realised, transformer action will take place between load and control winding and an alternating voltage injected into the control circuit from the load circuit.

A practical form of a simple magnetic amplifier circuit is illustrated in fig. 7.4. As can be seen the circuit contains two

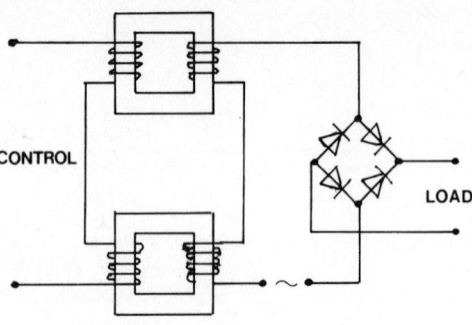

Figure 7.4

saturable reactors connected so that the control current establishes opposite fluxes in the two cores. This effectively neutralises transformer action between the control and load windings and results in one core saturating in one half cycle and the other core saturating in the other half cycle, control over the load voltage being obtained throughout the entire cycle. When one load winding is saturated its reactance falls to zero causing a high current to circulate in the load circuit and acting as a virtual short circuit to the other load winding. This winding then behaves much as a short-circuited transformer, its reactance also falling to a low level. The overall average load current/control current characteristic is shown in fig. 7.5.

This circuit as it stands is not sensitive to control current polarity, i.e. the average load current rises when the control current rises regardless of the direction of flow of the control current. (If the control current is reversed the same action is obtained as before but with the load windings saturating on the other half cycle to that on which they saturated previously.) To obtain polarity sensing the magnetic amplifier may be *biased* by the addition of a separate bias winding as shown in fig. 7.6. The bias current establishes a flux in the core whether or not there is a flux due to control or load winding currents and the magnetic amplifier now has an 'operating point' on its characteristic as shown in fig. 7.7. When the control winding current acts in one direction the control flux aids the bias flux and in the other direction the control flux opposes the bias flux, the net flux level changing accordingly.

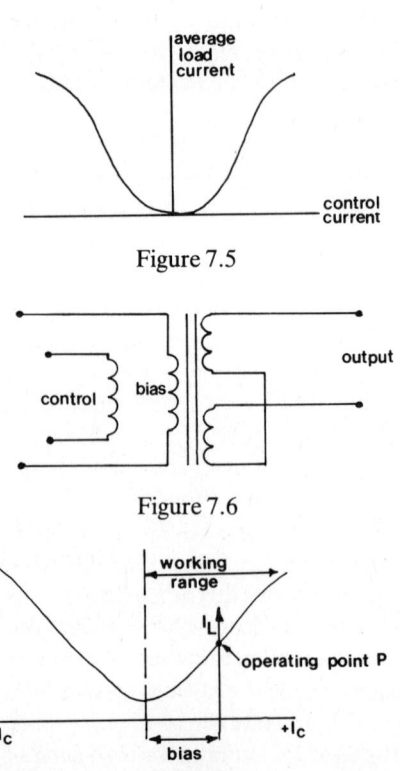

Figure 7.5

Figure 7.6

Figure 7.7

When no control current flows the bias flux and the flux produced by the load winding act together to give the specific value of average load current where the characteristic cuts the load current axis at point P. If the control current is now increased in one direction the average load current rises (right hand part of the graph of fig. 7.7) and if the control current is increased in the opposite direction the average load current falls (left hand part of the graph of fig. 7.7) The magnetic amplifier is now sensitive to control current direction.

Feedback

The amplification or gain of an amplifier is defined as the ratio of output to input, the quantities being voltage or current or power. The effective overall gain may be changed by taking part of the output and adding it to the input so as to either assist or oppose the input. The actual input received by the amplifier may thus be altered so that the ratio output to input (the input being the input *before the feedback signal is added*) may increase or decrease. If the input and thus the gain is reduced by feedback it is called *negative* and if the input and thus the gain is increased by feedback it is called *positive*.

Positive feedback is commonly applied to magnetic amplifiers, a typical circuit being shown in fig. 7.8. Here part of the output is

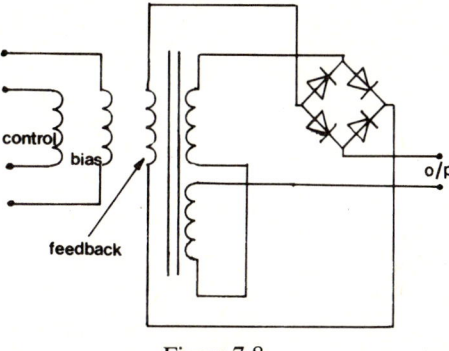

Figure 7.8

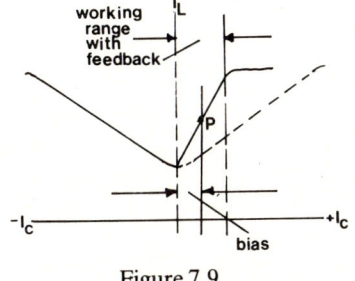

Figure 7.9

rectified and passed to a feedback winding connected so that an additional 'feedback flux' is set up in the core in addition to those already present. The overall effect is to increase the slope of the current characteristic, as shown in fig. 7.9, a larger change in average load current being obtained for the same change in control current. The gain is thereby increased and current gains of the order of one million are obtained without undue difficulty.

Advantages and disadvantages of the magnetic amplifier

The principal advantages of the magnetic amplifier are robustness, high reliability and efficiency and good stability with minimum drift at low frequencies. They are of course less compact and heavier than their semiconductor counterparts but nevertheless may be used to advantage as d.c. amplifiers in control and measurement applications.

General objective

The expected learning outcome is that the student explains the action of the thyristor, diac, triac and unijunction transistor.

Specific objectives The expected learning outcome is that the student:

9.1 *Sketches the characteristic of a thyristor.*
9.2 *Explains the action of a thyristor in terms of the two-transistor model.*
9.3 *Describes the action of the diac, triac and unijunction transistor.*
9.4 *Explains the operation of simple circuits which use thyristors, diacs, triac and unijunction transistors.*

The thyristor The thyristor or silicon controlled rectifier is shown schematically in fig. 7.10. The component makes use of the breakdown of a reverse biased p-n junction contained within two outer layers called the anode and cathode. When connected in a circuit as in fig. 7.10 it can

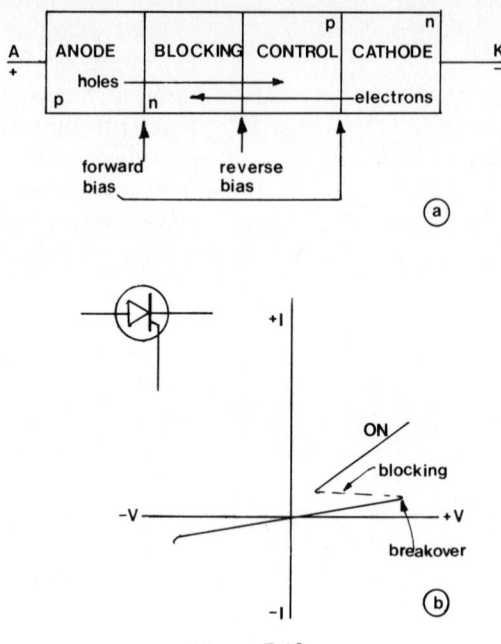

Figure 7.10

be seen that the anode and cathode junctions are forward biased but the control junction is reverse biased. Under these conditions the through current in the device is made up of minority carriers (electrons) in the p-type control layer moving to the anode and minority carriers (holes) in the n-type blocking layer moving to the cathode. The current is thus the saturation current of the reverse biased control junction. Additional carriers flow from the anode layer to the cathode (holes) and from the cathode layer to the anode (electrons), but the overall current is fairly small because of the blocking-layer/control layer junction reverse bias. As the diode p.d. is increased the through current rises but only slightly until a point is reached, called the *breakover point*, at which the reverse field is so strong that breakdown occurs at the control junction. With the anode and cathode now contributing heavily to the current, it rises rapidly and the device resistance drops sharply. The p.d. also drops and 'kicks back' to the conduction p.d. as shown in the characteristic of fig. 7.10b. The breakdown is reversible and the original state of the diode may be obtained by reducing the applied voltage until the current falls below the holding current as shown in the figure. The operating point then moves to the low conduction or off state.

The breakdown can be induced earlier by applying a positive potential to the control layer with respect to the cathode so that the reverse field is increased. The control layer is then known as the gate and an electron current flows into the control layer and hence to the anode. Once breakover has occurred, which it now does at a much lower voltage than before, the through current (anode to cathode) is independent of the gate–cathode p.d. and can only be reduced by reducing the anode–cathode p.d. as before. Thyristors are widely used in power supply and control circuits, some of which are given later in the chapter.

The triac

The thyristor once triggered conducts only in one direction, the two main terminals being called the anode and cathode respectively. The *triac* is a device based on similar principles and may be crudely considered to be two thyristors in one such that once triggered it conducts in either direction as determined by the respective polarity of the two main terminals. These terminals indicated by T1 and T2 in fig. 7.11 are not given the names anode and cathode since conduction is possible in either direction.

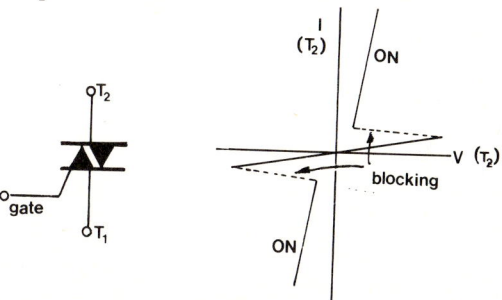

Figure 7.11

Triggering is achieved by applying an appropriate voltage to the gate, the polarity of this voltage with respect to either main terminal being unimportant, i.e. the device will trigger with either a positive or negative voltage though the level required may differ slightly for one polarity from that of the other polarity. A typical triac characteristic is shown in fig. 7.11.

The diac

Semiconductor diodes normally conduct in one direction only, conduction beginning once the junction p.d. set up by random carrier movement has been neutralised. This junction is quite small, usually of the order of tenths of a volt. The diac is a specially constructed diode which differs in two respects:

(a) it is bidirectional, conduction taking place in either direction;

(b) conduction does not begin until a much larger voltage, called the breakover voltage, is reached.

Typical values of breakover voltage are 30–40 V, typical currents being from fractions of a milliampere to one or more amperes. The device is used in triac gate firing circuits and helps to determine the triac firing point more accurately.

The unijunction transistor

The unijunction transistor is illustrated in cross-section in fig. 7.12. As can be seen, it is in fact a diode with two leads connected diametrically opposite to one another in one of the sections and a third lead connected to the end of the other section.

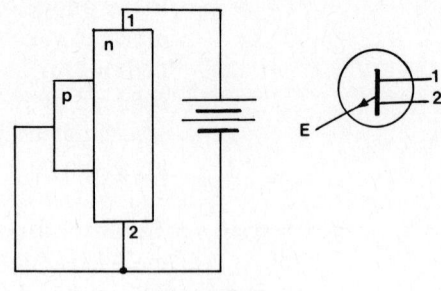

Figure 7.12

When connected as in the figure, a p.d. exists across the n-type material due to the battery. The potential at any point in the n-type material with respect to the battery negative pole is positive, the level depending on the position of the point relative to the negative pole. If the p-type material is connected to the battery negative pole the p-n junction is reverse biased, the level of bias changing as one proceeds down the junction from positive-pole connection to negative-pole connection. A sawtooth-generator circuit using this device is described later.

Gate firing circuits

A gate firing circuit for a thyristor must provide a pulse of sufficient duration to fire the thyristor and have its peak voltage occurring at the commencement of the pulse. The peak voltage, peak current and rise time must be within the limits specified in the device data.

Part of the basic circuit of a suitable trigger unit using a bipolar transistor is shown in fig. 7.13. Diodes D_1 and D_2 clip the input

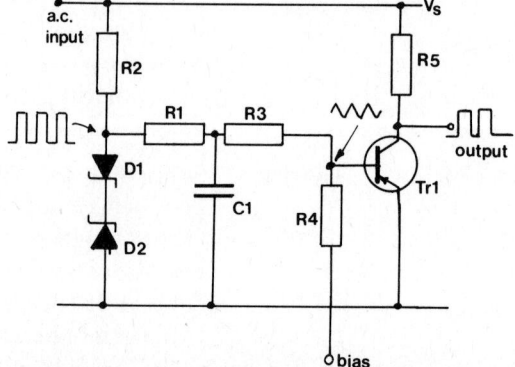

Figure 7.13

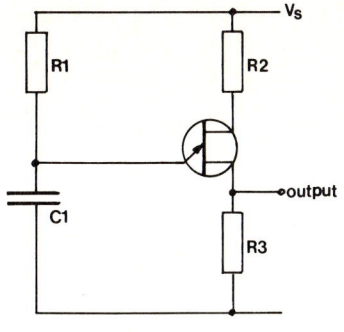

Figure 7.14

alternating voltage to the square wave shown. This square wave is passed thorugh an *RC* circuit to produce a sawtooth wave at the input of the transistor Tr_1. The pulse width may be adjusted by the application of a d.c. bias at the base or emitter of Tr_1. The transistor will then saturate only when the input waveform and the applied bias combine together to give sufficient forward bias. The d.c. bias may hold the transistor on, and the input may be used to turn it off or vice versa as required. A trigger circuit using a unijunction transistor is shown in fig. 7.14.

When no voltage is applied to the emitter of a unijunction transistor, only a small current flows across the base region from one supply connection to the other. At a certain value of emitter voltage, the emitter–base junction becomes forward biased and carrier injection occurs at the junction. Thus a large current flows through the base region and the effective resistance falls. The device has a negative-resistance characteristic once the trigger voltage is exceeded. In the circuit shown the capacitor charges with a time constant $C_1 R_1$ until the unijunction transistor triggers. The p.d. between base connection then falls rapidly. The capacitor p.d. falls as it discharges until it is low enough to again reverse bias the emitter–base junction and the process is repeated. The pulse output is positive if taken from the lower base resistor and may be used to trigger the thyristor directly. If d.c. isolation is required, the pulse may be fed to the thyristor via a transformer.

Thyristor and triac applications

Fig. 7.15 (a) and (b) show single-phase controllable bridge rectifiers. With conventional rectifiers the output voltage and ripple factor are fixed by the incoming supply and the rectifier characteristics. Once selected these are, of course, unchangeable during operation. The thyristor is a controlled rectifier and the point at which conduction begins (the conduction angle) is variable by means of the gate circuit. The average output, which determines the d.c. level, and the ripple content depend upon the conduction angle, and by suitably adjusting the firing circuits any level of ripple factor or d.c. output may be obtained.

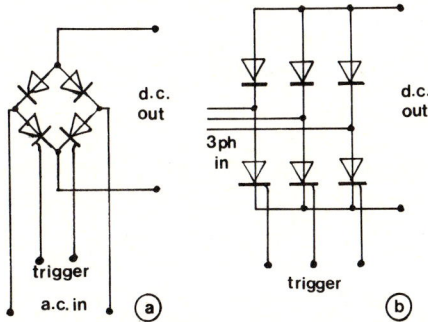

Figure 7.15

In mobile power systems, it is often required to generate an a.c. supply from a d.c. supply. One method of doing this is by using rotating machinery. The use of solid-state devices, however, is very often preferable owing to their inherent advantages of size, weight, reliability and economy. A thyristor invertor circuit is shown in fig. 7.16. In the circuit shown, thyristor A conducts whilst thyristor B is

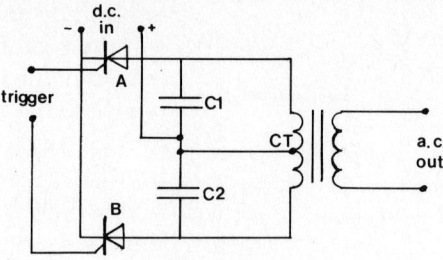

Figure 7.16

non-conducting and vice versa, the conduction periods being determined by the control circuit. When thyristor A conducts, current flows from the positive pole of the d.c. supply to the transformer centre tap then via the upper half of the primary winding to the conducting thyristor and then to the negative pole of the d.c. supply. Current flows from CT to A in the primary winding and a pulse is generated in the secondary or output winding. The capacitor C_1 and the transformer inductance combine to round off the pulse edges and an approximate semi-sinusoid is produced at the output. When thyristor A is off, thyristor B conducts and the lower half of the d.c. circuit conductors generating a similar approximate semi-sinusoid in the output. Since this part of the output is produced by current flowing from CT to B in the primary winding the pulse generated has the opposite polarity to that produced when thyristor A conducts. The resultant output is approximately sinusoidal.

The frequency of the output is determined by the conduction pulse rate, capacitors C_1 and C_2 and the transformer inductance and may be adjusted over a reasonable range. Modifications of the circuit can be used to generate very low frequency supplies having a high power output if so desired.

General objective The expected learning outcome is that the student applies the measuring instruments met so far to typical measurements.

Specific objectives The expected learning outcome is that the student:
8.1 Makes dB measurements
8.2 Measures the loading effects and frequency characteristics of simple measuring instruments.
8.3 Uses a CRO to demonstrate the presence of harmonics in various waveforms.
8.4 Uses a CRO to measure sine, square and pulse waveforms and phase differences.
8.5 Predicts the effect of distorted waveform on different detectors in electronic instruments.
8.6 Uses a commercial bridge for measuring inductance and Q-factor.
8.7 Calculates the total possible error, due to instruments.

dB measurements   A considerable number of tests carried out on instruments involve the comparison of two quantities or two magnitudes of the same quantity under different conditions. The comparison thus involves the use of a ratio. It is often convenient to express such a ratio using logarithms.

The most commonly used logarithmic ratio is the bel and its sub-unit the decibel which should be strictly applied to power levels (or voltage or current levels with the same load), but is often erroneously applied to other quantities.

A ratio of power levels expressed in bels is the logarithm to the base of 10 of the ratio. Thus for two power levels one of which is twice the other, for example, the ratio is log 2 bels, i.e. 0.3010 bels abbreviated 0.301 B. Alternatively, the ratio may be expressed as 0.5 since one level is half the other. In bels this is log 0.5 which, by definition, is $-1 + 0.6990$ or -0.3010 B. The same figure is obtained regardless of which level is taken as numerator but the sign changes. Thus it is said that the larger level in this case is 0.3010 B up on the smaller level or, alternatively, the smaller level is 0.3010 B down on the larger level.

A more convenient unit of ratio is the decibel, abbreviated dB, which is one-tenth of a bel. The ratio of two power levels expressed in decibels is thus *10 log (ratio) dB*.

Example 7.1   Express the following power level ratios in decibels: (a) 2 (b) 0.5

(a) 10 log 2 = 3.01 dB or 3.01 dB up.
(b) 10 log 0.5 = 10 × 1.6990
$$= 10 \times (-1 + 0.6990)$$
$$= -3.01 \text{ dB or } 3.01 \text{ dB down.}$$

If two power levels P_1 and P_2 are compared, where

$$P_1 = I_1{}^2 R_1$$

or

$$= \frac{V_1{}^2}{R_1}$$

and

$$P_2 = I_2{}^2 R_2$$

or

$$= \frac{V_2{}^2}{R_2}$$

then the ratio of P_1 to P_2 expressed in decibels is

$$10 \log P_1/P_2$$

which, from above, equals

$$10 \log \frac{I_1{}^2 R_1}{I_2{}^2 R_2}$$

$$= 10 \log \left(\frac{I_1}{I_2}\right)^2 \frac{R_1}{R_2}$$

$$= 20 \log \frac{I_1}{I_2} \text{ if } R_1 = R_2$$

or alternatively

$$10 \log \frac{P_1}{P_2} = 10 \log \frac{V_1^2}{R_1} \frac{R_2}{V_2^2}$$

$$= 10 \log \left(\frac{V_1}{V_2}\right)^2 \frac{R_2}{R_1}$$

$$= 20 \log \frac{V_1}{V_2} \text{ if } R_2 = R_1$$

Example 7.2   A signal generator is feeding a 1 mV signal into a 600 Ω load. Calculate the new input voltage if the generator attenuator is switched to −20 dB.
Since the load remains the same

$$-20 = 20 \log \frac{V_x}{10^{-3}}$$

where V_x is the new voltage input, so that

$$\log \frac{V_x}{10^{-3}} = -1$$

thus

$$\log \frac{V_x}{10^{-3}} = \bar{1}.0000$$

and

$$\frac{V_x}{10^{-3}} = 0.1$$

and

$$V_x = 0.1 \times 10^{-3} \text{ volts}$$
$$= 0.1 \text{ mV}$$

The new input voltage is 0.1 mV

Example 7.3   Calculate the voltage or current ratios corresponding to (a) 6 dB up, (b) 3 dB down.
The ratios will be the same for voltage or current since the equations have the same form. Thus

(a) $6 = 20 \log (\text{ratio})$

and

ratio = antilog 6/20
     = antilog 0.3
     = 2 (approximately)

Note that a 6 dB change in voltage or current corresponds to a 3 dB change in power.

(b)

$$-3 = 20 \log (\text{ratio})$$

and

ratio = antilog $(-3/20)$
      = antilog (-0.15)
      = antilog $\bar{1}.85$
      = 0.7079

i.e. the voltage or current is reduced to 0.7079 of the original for a 3 dB down adjustment. Notice that the wholly negative number -0.15 must be changed to the usual logarithmic form (i.e. a negative characteristic and positive mantissa) before tables can be used.

Occasionally the term decibel is applied to voltage or current levels where the load value changes but it should be noted that the attenuators or other controls calibrated in decibels are only accurate for voltage or current if the load remains constant.

Loading effects of instruments

Ideally the inclusion of measuring instruments into circuits should not affect the circuits so that the levels of voltage, current and power remain unaltered. In practice some loading effect does occur depending upon the nature and characteristics of the instrument. Loading effects are fully discussed in Chapter 8 of *Electrical and Electronic Principles 2*. Reference is also made in that chapter to the effect of frequency on measuring instruments, in particular that of the permanent-magnet, moving-coil variety.

Use of the c.r.o. to show distortion

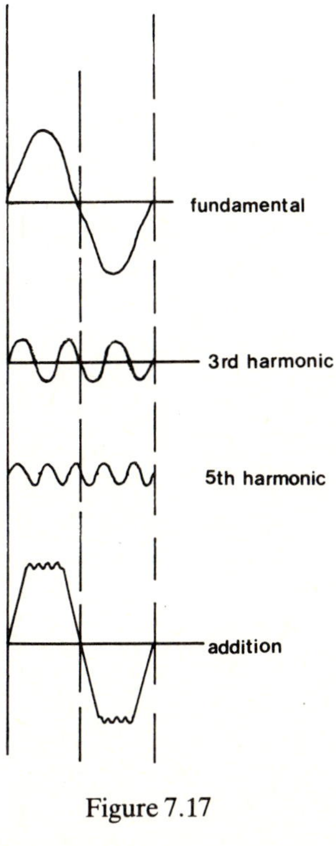

fundamental

3rd harmonic

5th harmonic

addition

Figure 7.17

The use of the cathode ray oscilloscope to display waveforms and measure voltage, current and frequency was described in Chapter 8 of *Electrical and Electronic Principles 2*. The instrument may also be used to indicate and measure phase difference between waveforms, angles of lead or lag showing clearly on the screen, and to indicate the presence of *harmonics* in waveforms.

Fig. 7.17 shows how an approximate square wave is made up of a number of component sine waves. In this example the frequency of each of the component sine waves is some simple multiple of the frequency of the largest component (called the fundamental frequency). The components are called harmonics of the fundamental frequency; a harmonic having a frequency equal to twice that of the fundamental is called a second harmonic; a component having a frequency equal to three times that of the fundamental is called the third harmonic, and so on. Fig. 7.17 shows that the sum of a fundamental frequency and its odd harmonics (three times the fundamental frequency, five times the fundamental frequency, seven times the fundamental frequency etc.) is a square wave. Note that the amplitude of each harmonic is progressively reduced as the order (that is, the number) of the harmonic is increased. It can be shown similarly that the sum of a fundamental frequency and its even harmonics yields a triangularly shaped wave.

Any complex waveform can be broken down into components. The shape of any signal wave can be changed as desired by passing the signal through suitable circuitry which affects one or more of the components more than the others. Also, it is possible to obtain a resultant complex waveform by summing components (not necessarily harmonics).

On occasion the shape of waveforms is changed undesirably by the circuits or instruments to which they are applied and on this occasion the change is called *distortion*. The c.r.o. is particularly useful for indicating the presence of distortion as shown in some typical distorted waveform displays as shown in fig. 7.18.

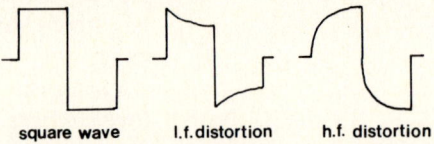

square wave   l.f.distortion   h.f. distortion

Figure 7.18

Distortion may introduce noticeable errors in measurement, one example being in the use of the moving coil instrument which is less accurate at high frequencies. Here the presence of higher order harmonics may reduce the accuracy of the reading of the waveform as a whole. Calibration of scale instruments is carried out using pure waveforms and again accuracy is impaired if the measured waveform has a shape substantially different from that used in calibration. If distortion is suspected the c.r.o. may be used to advantage to investigate the waveform being measured and appropriate allowance for inherent error can.be made.

Bridge measurements

The use of the Wheatstone bridge in the measurement of resistance was described in *Electrical and Electronic Principles 2*. A variety of other bridges have been devised for use with a.c. supplies for the measurement of inductance, capacitance, resistance and hence, since this depends upon the other three, the value of Q-factor.

Special considerations apply to a.c. bridges, particularly in the choice of supply and detector. The type of supply is determined by the frequency at which measurements are to be made. At low frequencies the mains supply or variable-frequency generator may be used, at audio frequencies and above an appropriate signal generator together with an amplifier, if necessary, to provide sufficient output is used. The detector may be a microammeter with rectifier, a cathode ray tube or at audio frequencies a pair of headphones, again with amplifier if necessary. Care must be taken to match the detector and bridge if maximum sensitivity is to be achieved. A.C. bridges are prone to errors due to external influences such as interference, radio frequency noise etc., and to the effects of internal stray reactances. The effects are largely determined by the frequency at which measurements are made and may be substantially reduced by careful screening of the bridge and the use of accurate components. For very accurate measurement, special circuits are used to reduce these various effects. Only the more common bridges are included below, and it should be pointed out that the analysis to give the balance equations is not included.

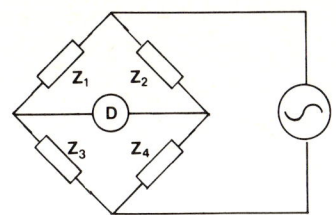

Figure 7.19

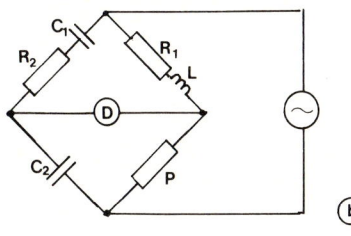

(a)

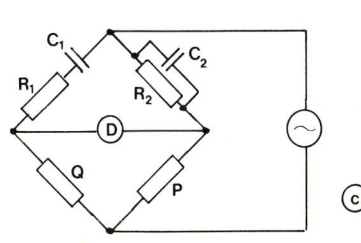

(b)

(c)

Figure 7.20

Errors in measurement

The general arrangement of an a.c. bridge is shown in fig. 7.19. For this circuit balance is obtained when

$$\frac{Z_1}{Z_4} = \frac{Z_2}{Z_3}$$

and

$$\Theta_1 - \Theta_4 = \Theta_2 - \Theta_3$$

where Z_1, Z_2, Z_3, Z_4 are the impedances shown and Θ_1, Θ_2, Θ_3, Θ_4 are the respective phase angles. Note that the impedance balance equation is similar to the Wheatstone equation for the d.c. case but in addition phase differences between arms must be equal.

Fig. 7.20 a, b and c show the Maxwell bridge, the Owen bridge and the Wien Bridge, respectively.

The Maxwell bridge shown is one of a number of possible arrangements. The balance equations are:

$$R = PQ/R_v$$
$$L = PQC$$

The circuit provides a useful means of measurement of coil resistance and inductance in one step. An alternative circuit is the Owen bridge for which the balance equations are:

$$L = PR_2\,RC_2 \qquad R_1 = PC_2/C_1$$

The Wien bridge is often used for frequency measurement and is also the basis of a well-known electronic oscillator circuit. The balance equations are:

$$\omega^2 = 1/R_1\,R_2\,C_1\,C_2$$

so that

$$f = \frac{1}{2\pi}\sqrt{\left(\frac{1}{R_1\,R_2\,C_1\,C_2}\right)}$$

The errors occurring in any observation or series of observations may be categorised into two types: random and systematic.

Random errors, as the name implies, are unpredictable as to size or nature ('low' or 'high'). They occur completely at random and as such for any set of observations of the same nominal value tend to be self-compensating, i.e. the probability is that 'high' errors compensate for 'low'. This is one of the reasons why the arithmetic mean of a set of observations is the best value (in the sense that it is closest to the true value), since the compensating ability has full opportunity to work in this case.

Systematic errors are predictable errors and their cause may be determined and eliminated. They occur regularly, i.e. with each observation, and as such tend to be cumulative. They may be sub-categorised into observer error, instrument error and natural error. The names are fairly self-explanatory in that observer error is due to some characteristic of the person making the observations, for example a tendency to read 'high' or 'low' when estimating the position of a pointer between calibrations. Observer error is usually compensated by a change of observer and making several sets of observations. Instrument error is due to some defect within the instrument used for measuring. In the case of instruments undergoing tests it is this error which is being sought. A detailed discussion of instrument error for the more common instruments used under various conditions is contained in *Electrical and Electronic Principles 2*. Natural error is due to natural or physical phenomena such as expansion of materials when heated, the influence of atmospheric pressure or content, etc. Once found and distinguished by cause, compensation may be made for this type of error.

Summary

A magnetic amplifier is able to control high levels of power by small signals. The principle of operation is the variation of the inductance of iron cored coils by altering the flux level within the coil cores. A practical magnetic amplifier has a load winding, a control winding (to which the input signal is applied), a bias winding (to enable the amplifier to distinguish signal polarity) and a feedback winding (so that higher gains may be obtained by the application of positive feedback). The main advantages of the magnetic amplifier are robustness, high reliability and efficiency and good stability with minimum low frequency drift.

The thyristor or silicon controlled rectifier has three connections, anode, cathode and gate. Through current is controlled by the gate potential which determines the point at which the thyristor conducts when a changing voltage is applied between anode and cathode. Once conducting the thyristor can be switched into non-conduction only by reducing the anode–cathode voltage. Unlike the thyristor the triac conducts in both directions, the point of triggering being controlled by the gate. The diac is a special kind of solid-state diode which conducts once a particular value of anode–cathode voltage, the breakover voltage, is reached. The unijunction transistor has somewhat similar characteristics but works on a different principle, the control electrode being connected internally to the cathode, in effect making the device a two-connection component similar to a diode. All these solid-state devices are used extensively in power control circuits.

The decibel is a means of comparing two power levels, the comparison being expressed as

$$10 \log P_2/P_1 \text{ dB}$$

where P_2, P_1 are the power levels. Provided that they are applied to the same values of resistance, voltage and current levels may also be compared as

$$20 \log V_2/V_1 \text{ or } 20 \log I_2/I_1$$

where V_2, V_1, I_2, I_1 are the voltage and current levels respectively.

The bridge principle used in the basic Wheatstone bridge may be applied to a.c measurements and there are a large number of possible a.c. bridge circuits. All of them consist of the bridge connection, a detector and an a.c. source. The balance equations concern both voltage and phase and contain component values as determined by the bridge type.

Possible marks

SELF-ASSESSMENT EXERCISE 7

1. List three advantages of the magnetic amplifier. (3)

2. Give two types of feedback and indicate the effect on amplifier gain when they are applied. (3)

3. Draw the current/voltage characteristic of a thyristor. (3)

4. What is the principal difference between a thyristor and a triac? (3)

5. What is a diac? (3)

6. Draw a typical current/voltage characteristic of a triac. (5)

7. What is the purpose of:
   (a) bias winding;
   (b) control winding in a magnetic amplifier? (5)

8. A 5 V supply is attenuated by −5 dB. Calculate the new voltage output. (5)

9. Draw the circuit diagram of a typical practical magnetic amplifier showing all necessary windings. Describe the function of each winding including in your description a typical magnetic amplifier control characteristic. (14)

10. Briefly describe the principle of operation of the following:
    (a) diac;
    (b) unijunction transistor;
    (c) thyristor. (14)

11. (a) With the aid of waveform diagrams explain what is meant by harmonics of a regularly recurring cyclic waveform. Show how harmonics may be summed to give an overall composite waveform different from that of the fundamental waveform.
    (b) Sketch a square wave which has been subjected to poor low frequency response in an electronic circuit. (14)

12. Describe the fundamental principles of an a.c. bridge and the actual circuit of any *one* of the bridges given in the text. Include the balance equations in your answer. (14)

13. Define the bel and decibel in terms of power, voltage and current levels. The voltage and current relationships should be derived and the special condition fully explained.
    A signal generator supplies 100 mV into a 600 Ω load. Calculate the supply voltage at the following settings of the attenuator switch:
    (a) 3 dB;
    (b) 6 dB;
    (c) 20 dB. (14)

Answers

Marks

SELF-ASSESSMENT EXERCISE 7

1. Any three of those given in the text. (1 each)

2. Positive feedback increases gain; negative feedback reduces it. (1½ each)

3. Axes must be labelled and all relevant points shown. See fig. 7.10b. (3)

4. A thyristor conducts in one direction only; a triac conducts in both. (3)

5. See text. Breakover voltage must be mentioned. (3)

6. See fig. 7.11. *All* relevant points must be shown for full marks. (5)

7. See text.

(Bias winding 2½)
(Control winding 2½)

8.
$$-5 = 20 \log \frac{V_x}{5}$$ (1)

$$\log \frac{V_x}{5} = -0.25$$

$$= \bar{1}.75$$ (1)
  (1)

$$\frac{V_x}{5} = 0.5623$$ (1)

and $V_x = 2.81$ V (1)

9. See fig. 7.8, 7.7 and text. Deduct 1 mark for each item omitted.

(Circuit diagram 4)
(Function of each winding (four) 1½ each)
(Characteristic, to include working range and operating point etc. 4)

10. As text. Full details to be included for full marks. Deduct 1 mark per item omitted.

(a 4)
(b 4)
(c 6)

11. (a) See fig. 7.17 and text.

(Diagram 6)
(Explanation 4)

(b) See fig. 7.18

(Sketch 4)

12. See fig. 7.19. Bridge is based on balancing arms in impedance and phase. (4)
Actual bridge circuit. (4)
Balance equations. (6) (3 each)

13. Definition. (3)
Derivation. (3)
Voltage and current relationships only correct if resistance remains the same. (2)

(a)
$$-3 = 20 \log \frac{V_x}{100} \quad (V_x \text{ in mV})$$

hence $V_x = 70.79$ mV (2)

(b) $V_x = 50.11$ mV (2)

(c) $V_x = 10$ mV (2)

Index